Politics and Aesthetics of Creativity
City, Culture, and Space in East Asia

Edited by
Lu Pan
Heung Wah Wong and
Karin Ling-fung Chau

Politics and Aesthetics of Creativity
City, Culture, and Space in East Asia
Edited by Lu Pan, Heung Wah Wong, and Karin Ling-fung Chau
Copyright © 2015

Distributed by Transaction Publishers
10 Corporate Place South, Suite 102
Piscataway, NJ 08854

All rights reserved. Exclusive English language rights are licensed to Bridge21 Publications, LLC. No part of this book may be used or reproduced in any matter whatsoever without written permission from the publisher except in the case of brief quotations embodied in critical articles and reviews.

For information contact Bridge21 Publications, LLC, 11111 Santa Monica Blvd, Suite 220, Los Angeles, CA 90025.

Published in the United States
Cover Design by Chi-Wai Li
Copyedited by Valerie Spanswick
ISBN 978-1-62643-016-7 Paperback
      978-1-62643-017-4 eBook

Table of Contents

List of Contributors 7

Preface 11

Chapter One: The Mall as Political: The Urban Interweaving of Bangkok's Past and Present 17
    Trude Renwick

Chapter Two: Small-scale and Bottom-up: Innovating Creativity in Shanghai's City Center Quartiers 55
    Ying Zhou

Chapter Three: In Praise of the "Coffin": Urban Sociality in the Japanese Capsule Hotels 93
    Non Arkaraprasertkul

Chapter Four: Rediscovering the Japanese Houses in Taiwan: A Contest between Postcolonial Inhabitants and the Creative City Regime 119
    Shu-Mei Huang

Chapter Five: The Kitschy, the *Shanzhai*, and the Ugly: Creating Architectural Utopia in Contemporary Chinese Cities 153
    Lu Pan

Chapter Six: (In)Dependence, Industry, and Self-Organization: Narratives of Alternative Art Spaces in Greater China 183
    Elaine W. Ho

Chapter Seven: On the Finding, Producing, and Losing of Creative Space: Hong Kong's Hidden Agenda 219
   Tobias Zuser

Chapter Eight: The Tourist Gaze, the Invisible Cities: Cultural Heritage and Tourism in Taiwan..245
   Yi-Chieh Lin

List of Contributors

Non Arkaraprasertkul is a Global Postdoctoral Fellow at New York University Shanghai. He was trained in architecture and urban design at the Massachusetts Institute of Technology (MIT), in history and modern Chinese studies at the University of Oxford, and in anthropology at Harvard. His research interests lie at the crossroads of transdisciplinary research between architecture and the social sciences. He was also a Fulbright Scholar and Asian Cultural Council Fellow (2005–2007), visiting lecturer in architecture and urban design at MIT (2007–2008), Harvard-Yenching Doctoral Scholar (2008–2011), adjunct professor in modern Chinese history at Lesley University (2012), Distinguished (Visiting) Gibbons Professor of Architecture at the University of South Florida (2012–2013), and Chinese Scholarship Council (CSC) Visiting Scholar as well as Fellow at the International Center for Studies of Chinese Civilization, Fudan University, Shanghai (2013 – present).

Elaine W. Ho (HK/USA) works between the realms of time-based art, urban practice, and design, using multiple vocabularies to explore the micropolitics of everyday life. Often working collaboratively, her work begins foremost from forms of documentary-making in an attempt to describe, touch upon, and make small gestures toward the alter-possibilities of an intimate, networked production. She is the initiator of the artist-run project space HomeShop and a frequent contributor at www.iwishicoulddescribeittoyoubetter.net.

Shu-Mei Huang received her PhD from the Program in Built Environment at the University of Washington. She is curently teaching at the Graduate Institute of Building and Planning, National Taiwan University. Her research interests include postcolonial urbanism, critical heritage studies, and care transnationalization; recent research centers on representations of colonial prisons as heritage of incarceration in Asia. Her manuscript titled "Urbanizing Carescapes of Hong Kong: Two Systems, One City," based on her dissertation, is now on press with Lexington Books (Lanham).

Yi-Chieh (Jessica) Lin is Assistant Professor at College of Communication in National Chengchi University. She received PhD degree in Anthropology and A.M. in Regional Studies- East Asia both from Harvard University. Her research interests lie in cross-cultural communication, cultural heritage issues in East Asia and risk communication. She is author of "Fake Stuff: China and the Rise of Counterfeit Goods".

Lu Pan received her PhD in Comparative Literature from the University of Hong Kong. Her current research interests include visual culture, urban space, war memory, and theories of aesthetics. Dr. Pan did her research as a Visiting Fellow at Berlin Technical University and Harvard Yenching Institute. She is author of two upcoming monographs, "Aestheticizing Public Space: Street Visual Politics in East Asian Cities" (Bristol: Intellect) and "In-Visible Parlimpsest: Memory, Space and Modernity in Berlin and Shanghai" (Bern: Peter Lang). Since August 2015, she works as Assistant Professor at Dept. of Chinese Culture, Hong Kong Polytechnic University.

Trude Renwick is a PhD candidate in Architecture studying History, Theory and Society with a concentration in Anthropology in the College of Environmental Design at the University of California, Berkeley. Her current research focuses on the market, examining malls, "informal", and "abandoned" urban spaces and their interconnections within the urban fabric of Bangkok. A Wisconsin native, she received her BA from Brandeis University in Anthropology and her MA in Design Studies from the Harvard Graduate School of Design focused on the History and Philosophy of Design.

Ying Zhou is an architect who is researching city-center transformation processes in Shanghai between 1992-2012 with a focus on how the spatial production system has evolved between the aspiration to be global and the residual local frameworks. She has taught at the ETH Studio Basel on the research studios for Kolkata, Damascus, and Cairo, as well as at the Singapore-ETH Centre on urban designs for the Rochor area in Singapore. She has published in Critical Planning, Urban China, and Monu; and her work has been exhibited at the Swiss Architecture Museum in Basel, the Haus der Kunst in Munich, and the Shenzhen/Hong Kong Biennale. Ms. Ying trained at Princeton (B.S.E.) and Harvard (M.Arch) and was a Fulbright Fellow at the University of Stuttgart. She is currently at the ETHZ, The Swiss Federal Institute of Technology (zhou@arch.ethz.ch).

Tobias Zuser is a PhD Candidate at the Department of Humanities and Creative Writing at Hong Kong Baptist University, and received his MPhil in Cultural Studies from Lingnan University, Hong Kong. Originally trained in the field of Culture, Sports, and Event Management in Austria, Mr. Zuser has worked for more than three years as an arts and event manager in Berlin and Beijing before returning to academia. Based on his experience as a practitioner, his essay "How the Cultural Sector Works in China" was published in a bilingual handbook by the European Union National Institutes for Culture (EUNIC) in 2011. His current research interests include cultural policy, creative industries, alternative culture, urban redevelopment, sports policy and football culture, with a strong focus on Hong Kong and China.

Preface
Lu Pan, Heung Wah Wong, and Karin Ling-fung Chau

This collection of articles was inspired by an international conference titled "Politics of Creative Industries: Critical Reflections" organized by the Global Creative Industries Programme at the University of Hong Kong in March 2013. We had interesting conversations and debates on the relationship among politics, aesthetics and creativity at the conference. One of the participants and, later, editors of the book, Lu Pan, went ahead to collect and organize papers related to the themes of this edited volume: the dynamics between the production, use and interpretation of space and the social transformation of the East Asian region, and the role of politics and state in shaping that dynamics.

Traditionally and, in fact, ritualistically, an introduction to an academic volume starts with a discussion of the theoretical framework of the book which is to be followed by a summary of each chapter. In this introduction, we want to experiment with something new. Instead of simply summarizing each chapter and restricting the imagination of readers to a specific theoretical framework, we attempt to spell out some common theoretical and empirical implications of the chapters that we consider relevant and important to the study of creative industries specifically and the study of culture and society in general. We invite our readers to engage in the process together and judge if our venture is successful.

All of the chapters in the volume are concerned about space and architecture in the urban setting of East Asia. Chapter 1 discusses a central mall area in Bangkok; Chapters 2, 6, and 7 examine creative clusters and alternative art space in Greater China; Chapter 3 focuses on the capsule hotels in Tokyo; Chapters 4 and 8 survey heritage sites in Taiwan whereas Chapter 5 analyzes contemporary architectural practices in China.

A central theme connects all these chapters with diverse subjects of study: an endeavor to explore and understand space and architecture as a model *of* and *for* society. On the one hand, specific space and architectural sites are examined in the volume as texts which reflect social reality. In other words, the space is constituted by hidden meaning embedded in the social structure and context. On the other hand, the same space and architectural sites are contested arenas shaping and being shaped by the economic structure, political institutions and cultural logic of our society. The complexity of the process is identified, scrutinized and evaluated from various perspectives in the volume.

Guided by this underlying theme, we are able to further infer from the chapters two major implications to the studies of creative industries and social transformation in East Asia. The first one is related to the role of the State in the development of creative industries, and the second delves into the intricate relationship between the politics of space, the local social context and cultural logic.

In particular, the findings of Chapters 2, 6, 7 and 8 lead us to ponder the impact and effectiveness of State planning on the development of creative industries and the process of cultural revitalization. As shown in these chapters, the governments of China, Taiwan and Hong Kong all play an active role in 'curating' and planning, if not standardizing, the development of 'creative clusters' (Chapters 2, 7, 8) and the art industry (Chapter 6) in the region.

Discrepancies have been found, however, between the official discourse and policy and the actual reality of local development. For instance, Chapter 2 surveys the development of non-State-designated creative clusters in Shanghai initiated by creative entrepreneurs, designers and residents, while Chapter 7 examines the difficulties encountered by an illegal music venue in an industrial district in Hong Kong. Chapter 6, on the other hand, attempts to present us a close-up picture of independent art organizations which do not succumb to the dominant State paradigm and industry regime. Chapter 8 outlines the process of regenerating and rebranding cultural heritage into cultural and creative products in Taichung and warns about the danger of gentrification in the region.

The lesson that we can draw from the above studies is that, in reality, dual forces are at play in the creative industries in the Greater China region. On the one hand, the State implements an official blueprint for creative industries and produces, accordingly, an official discourse of creative industries, legitimizing the official categorization and rewarding the official targets of development.

On the other hand, alternative and grassroot models of creative industries spring up along with and in response to the official paradigm.

The dual forces of the State and the 'alternative' are constantly interacting and reshaping each other. This dynamic, as elaborately illustrated in the case studies of Chapters 2, 6, 7, 8, prompts us to explore the empirical question of the prospect of the creative industries. What should be the role of the State in the development of creative industries? Is a bottom-up model preferable to a top-down approach?

These questions are particularly relevant to the study of the creative industries in East Asia. The global surge of 'Hallyu' (Korean Wave) in the recent years, for example, attracts heated interest in comparing the State-led model of Korea with the market-oriented approach of Japan.

Before one embarks on the journey to look and shop for effective cultural policies, we want to emphasize one important point. One should bear in mind that the 'state' and the 'alternative' or 'local' that we mention here are not homogenous entities. The process of policy-making and implementation is constituted by heterogeneous voices and agencies; on the other hand, local organizations and actors are also marked by internal contradictions and conflicts, as demonstrated in Chapter 6. In other words, we have to look into the relations and mediations among different actors across different levels instead of essentializing the non-existent dichotomy between 'State' and 'Alternative'.

In addition to giving an insight into the relationship between the State and the formation of creative clusters and art spaces, this volume also shed lights on the broader question of the politics of space. Chapters 1, 3, 4 and 5, in particular, reveal that the making, use and interpretation of architecture and urban planning are ordered, if not prescribed, by local cultural logic and social context.

Chapter 1, for instance, suggests how the concept of vertical holiness in Thai society is embedded in the spatial arrangement of a central mall area in Bangkok, a space which manifests the social stratification of local society and became a locus of protest. Chapter 3 attempts to explain the phenomenon of the proliferation of capsule hotels in the contemporary Japanese context. It argues that guests do not choose to stay at the capsule hotel only out of necessity; they are actually attracted to seek social sanctuary there. This finding has to be understood in terms of the form of social alienation resulting from the commuting network and specific salarymen culture in Japan.

Chapter 4 challenges the dominant 'creative city regime' which depoliticizes and aestheticizes the colonial history of Taiwan through describing and discussing the surging interest in preserving and revitalizing Japanese Houses in Taiwanese society. The attempt to depoliticize the colonial history of Taiwan, we assert, is related to the broader nation-building project of Taiwan. The two dominant political parties in Taiwan, the Chinese Nationalist Party (KMT) and the Democratic Progressive Party (DPP), adopt different nation-building approaches. The KMT implements the policy of 'sinicization' while the DDP attempts to build a new Taiwanese identity by *culturally* distancing Taiwan from Mainland China. The DDP formula of the new Taiwanese identity basically is to recognize or even encourage the ethnic awareness of various Chinese ethnic groups and aboriginal groups. In addition, the DDP government in the early 2000s emphasized the Japanese cultural legacies of local Taiwanese culture. All of this aimed to demonstrate that while Chinese culture is part of the local Taiwanese culture, the latter consists of many other cultures, including Japanese culture, that never were part of the Chinese culture.

The logical consequence, as the DDP government would like to imply, is that according to the logic embedded in the Chinese native concept of "Guo-jia" (nation-state) that members of a nation-state should inherit the same history, speak the same language, belong to the same race, and share the same culture, since Taiwan and China have different national cultures, they should be two different nation-states. That is to say, cultural separation will lead to political separation! We therefore argue that the aestheticization of the Japanese Houses in Taiwan discussed in Chapter 4 is to be better understood against the local Taiwanese context in which the colonial Japanese legacies are revalued to symbolically mark the demarcation of Taiwan from Mainland China.

Lastly, Chapter 5 demonstrates how the contemporary architectural practices of China should be understood in terms of the local social and political context. It suggests that the proliferation of miniatures and 'Shanzhai' buildings is a renewed version of the Chinese Communist utopia, and the public debate on the 'ugly' buildings conceals the expression of public discontent by manifesting it in a 'safe' and sarcastic way.

All these studies remind us that we have to contextualize our understanding of the architecture, space and urban setting of each respective society.

To understand the politics of space and study its relationship with social transformation is a very demanding, yet rewarding task. This volume contributes to the field of study by providing lucid and specific case studies in East Asia and challenging dominant official discourses and practices. The studies here also never fail to remind us that we have to pay close attention to the complex dynamics between local context and global influence in order to formulate a more accurate interpretation of the spatial world. We hope that the volume will inspire scholars, cultural professionals and the public to take a new perspective to perceive and interpret spatial visuals in our own society vis-à-vis others.

CHAPTER ONE:

The Mall as Political:
The Urban Interweaving of Bangkok's Past and Present
Trude Renwick

Figure 1

Introduction

In the spring of 2010, Red Shirt protesters took over the Ratchaprasong Intersection in the urban economic core of Bangkok. This was just one major protest of many between the Red and Yellow Shirt movements in Thailand. The Red Shirts, a group organized by the National United Front of Democracy against Dictatorship (UDD), called for Prime Minister Abhisit Vijiajiva to hold early elections. As the leader of the Democrat Party, Vijiajiva and his government came to power during the 2008 judicial coup that banned the Palang Prachachon Party, led by Thaksin Shinawatra. These protests resulted in the burning of Central World Mall, the Siam Theatre, and surrounding low-density shops. The violence climaxed in a clash between the Thai military and the protestors that left about ninety-one protestors dead and ended the occupation.

The burning of this luxury mall and occupation of this portion of the city can be interpreted as a class conflict, rooted in local social structures. Yet the location or placement of these protests, in the shadow of a huge number of malls that serve as monuments to neoliberal economic policies, seems to suggest otherwise. Further, some architects dismiss or interpret these global, everyday spaces of popular culture like malls as anonymous and inherently "uncreative." In his essay "Junkspace," Rem Koolhaas states that forms like the airport and the mall are not form but "Proliferation.... Regurgitation is the new creativity; instead of creation, we honor, cherish, and embrace manipulation.... Superstrings of graphics, transplanted emblems of franchise and sparkling infrastructure of light, LEDs, and video describe an authorless world beyond anyone's claim, always unique, utterly unpredictable, yet intensely familiar" (2002: 177).

Although seemingly "authorless," the creativity that makes up the Siam Square area is not solely produced from "regurgitation" or "emblems of franchise." I agree with anthropologists and historians who argue that this overarching definition from the standpoint of the architect obscures the subtexts and meanings of form that are present surrounding the Ratchaprasong and Siam Square area (Low, 1995). Setha Low, in her book *The Politics of Public Space*, describes a relationship between politics and aesthetics asserted by other social theorists like Paul Rabinow and Michel

Foucault: "Political implications lie at the root of all aesthetic sensibilities, and certainly the design of an urban space reflects the political agency of the state. In this sense, architecture and urban design contribute to the dominance of one group over others" (Low, 1995: 748).[1] Aihwa Ong, in her analysis of Rem Koolhaas's China Central Television (CCTV) tower in Beijing, also attempts to move beyond the designation of such buildings as simply supranational consequences of globalization. Ong specifically addresses the ways in which this building becomes a symbol of the state's attempts to control the flow and dispersal of media, using the act of "hyperbuilding" to define this process; a term borrowed from Koolhaas, but redefined as the "process of building to project urban profiles" and the act of staging sovereign power through a physical landmark (Ong 2011, 207). The luxury malls connected to the Siam Malls and Central World Malls in the Siam Square area represents this staging of sovereign power described by Ong and have come to represent internal class tensions within the nation like many monumental spaces and urban projects across the city's history.

The Siam Square and Ratchaprasong area is, without a doubt, symbolic of the global wealth that plays a significant role in recent political and class conflicts, as well as being one of many monumental spaces and urban projects throughout the city's history developed by powerful groups that incorporate Western architectural forms (monarchy, military, and the capitalist elite). In this essay I do not solely focus on how this urban space of the mall contributes to the "dominance of one group over others," but I examine how seemingly distinct global and local spaces are intertwined in this area.

While on a monumental scale, the malls that make up the Siam Square and Ratchaprasong area incorporate local conceptions of power and spatial hierarchies, low classed spaces, street markets and the subversive political practices facilitating them, are also encouraged by and inserted into the global and supposedly disconnected space of the mall. Therefore, these spaces are connected to Margaret Crawford's concept of "everyday urbanism." In *Everyday Urbanism,* Crawford states, "We don't regard everyday space as a major aesthetic problem like the New Urbanists or call it "Junkspace" like Rem Koolhaas, but see it as a zone of possibility and potential transformation" (2005: 19). The hybrid and interwoven spaces that make up Siam Square are not only reinforcing local power structures through preexisting methods of incorporating Western forms, but this area dominated by malls serves as "a zone of possibility and potential transformation" by expanding this tradition of hybridity by incorporating low classed spaces. The physical relationship of the "anonymous" mall relative to its "local" surroundings and use in the everyday as well as the historical relationship between monuments of global culture, urban/political development, and protests in Bangkok reveal that Siam Square, although a part of a global, monumental mall culture visible in many cities across the world, is an inherently creative space in its own right.

The creativity present in this area does not solely reside in the plans and aerial mappings made by architects, real estate developers, and urban planners, but by spaces and moments created in the interaction and tying of informal and formal urban space. Thailand's rich history of hybridity within its capital city and politics alike actually moves beyond the "glocal" or binary between global and local. The careful—and often conscious—stitching together of these seemingly distinct layers of urban fabric occurs through institutions like religion, monarchy, military and politics. Not a new phenomenon in the context of Thailand, this interaction reveals spaces of everyday tension and balancing that emerge between classes, which then explode during events like the 2006 and 2010 protests. The protests and the occupation of this area are not a consequence of the form so much as a continuation and transformation of preexisting local class tensions that lie latent within this symbolic national space.

The street market, spiritual spaces and infrastructure within Siam Square are the arteries of daily life that simultaneously reassert and transgress social hierarchies. It is in this area that the "uncreative" malls, which dominate it architecturally, become connected to the tensions that are amplified in the context of the protest. Therefore, following, Crawford's as well as Low's and Ong's argument about the interconnected nature of politics and architectural and urban form, not only do creativity and political tension exist in this area via its role as a monument within the city around which protests occur but also in its use as an everyday space.

The following section begins by mapping a history of monumental spaces across Bangkok that simultaneously evoke Western architectural traditions and local conceptions of power and space. The second section, *Mall as Monument, Mall as Everyday*, examines how the space of the mall that dominates the

Siam Square and Ratchaprasong area serves as a continuum of this hybrid monumental space found within Thailand's capital. While infrastructural and spiritual spaces have a long history of incorporation into monumental spaces across the city, I use the example of the street market to demonstrate how this monumental space dominated by the mall is unique in that it also incorporates rather than juxtaposing itself with "lower" classed spaces. In other words, malls and their operators do not solely serve as monumental spaces, but also as everyday spaces; rather than solely removing the form of the street market they incorporate it in form and practice, all facilitated by their focus on maximizing profit. A fine balance is achieved in this area, not simply through the regulation of space but also by the incorporation of lower-classed spaces like the street market. I conclude by returning to this portion of the city as a space of protest, describing how latent tensions between classes explode. As the form of the "informal" or "lower" classed street market takes over this area through the "unwanted" or out of place Red Shirt protesters in 2010, the spiritual and infrastructural spaces described throughout this chapter become exaggerated in their roles as protective and demarcating entities within the area. It is in the hybrid space of play visible through the space of the mall in Bangkok, between global and local and high and low practices, that becomes exaggerated and unbalanced in the context of the protests. Ultimately, creativity, as described in this chapter, is not manifest in the monumental space that Siam Square and the Ratchaprasong intersection symbolize alone, but in its ability to take on lower spatial practices that complicate our understanding of the mall as a monolithic form that is both disconnected from local conceptions of power and low classed spaces of society.

History and Hybridity

Thailand's rich history of creative ordering encompasses both local and Western forms of politics, which is reflected in the physical development of the city, including the Ratchaprasong intersection. This hybrid creative ordering is arguably linked to Thailand's "crypto-colonial" past (Herzfeld, 2002). In many ways, Thailand has been greatly influenced by Western conceptions of aesthetics and order; however, unlike many developing countries, present-day Thailand emerges from neither a history as colonized nor colonizer. Michael Herzfeld defines this condition in his piece *The Absent Presence* as a "crypto-colony": "Buffer zones between the colonized lands and those as yet untamed were compelled to acquire their political independence at the expense of massive economic dependence, this relationship being articulated in the iconic guise of aggressively national culture fashioned to suit foreign models" (2002: 901).

Thailand's history of hybridity was not spurred so much by an erasure and replacement of Thai culture, but it was a selective accentuation of its own unique "civilization." This was a conscious decision to make certain concessions to Western powers like France and England in order to maintain autonomy. The presence of Western powers in Thailand dates back to the seventeenth century; surrounded by what became the colonies of Burma, Cambodia, Malaysia, and Laos, land and economic concessions were arguably essential for Thailand to remain autonomous politically. Therefore, while technically an independent nation, Thailand's emphasis on hybridity is visible both in the urban environment and in politics. Bangkok, as the ultimate symbolic center of the nation, received the brunt of this relationship with Western colonizers, reasserting in this core of the nation the "civilized" nature of those in power; it was a "Western-style state, governed—and analyzed—according to a positivistic understanding of the meaning of data and order and yoked to Western-derived ideas of a progressivist model of the past" (Herzfeld, 2012: 1). While this Western-style state that developed in Thailand is "yoked" to a "progressivist" and all-encompassing notion of governance, there is a simultaneous tension produced in its interaction with a more flexible and a geographically varied potency of state power historically present. Its hybrid of Western and Thai conceptions of politics does not function as a unified whole in Thailand; thus, various social and economic inequalities persist and, at times, explode.

In her book *Materializing Thailand*, Penny Van Esterik describes the many ways in which the Thai state plays with the "surface exoticism of locality and global transnational processes" (Esterik, 2000: 4). One word that can be used to describe the regulation of the ways in which different classes interact within Thailand, as well as the selective ways in which western political and spatial practices are inserted into the urban fabric, is *kalatesa*. *Kalatesa* is a term meaning the proper ordering of time and space in Thailand; it defines a number of practices, including the proper ways in which one pays respect to people according to their rank. *Kalatesa* originally appeared as the way in which a balance that is achieved within the body to best maintain health, but according to Esterik "It may have been an important means of social control… rural peasants and urban poor would have to exhibit knowledge of kalatesa in order to interact effectively with government officials" (2000: 40). One way in which manners and power is written into and defined in everyday space is the correlation of different levels of holiness to height. The head as the highest and the foot as the lowest parts of the body are defined and the most and least holy parts of the body. One must be care to not point or touch others with one's feet and treat books with special respect as they are tied to the mind and acquisition of knowledge. The practice of such beliefs about high and low spaces moves beyond the body and dictates movement within the city itself. When the royal family occupies a glass skyscraper it is expected that those in surrounding buildings move to a lower level out of respect. This also occurs on the massive elevated highway systems; one must either pull to the side of the highway, or one is prevented from entering the highway when the royal family and especially the King use it.

One has *phit kalatesa* when "the coming together of time and space and relationships is not quite right" which can result from either not knowing the proper type of behavior or by refusing to operate within the system of *kalatesa* (2000: 41). There is an appropriate time and place for expressing opinions and one should certainly not upset or offend others through such inappropriate behavior as it violates *kalatesa* and results in the loss of respect and face. In other words, other than setting the everyday ways in which one must display respect for elders, the monarchy, and those of higher social rank from your own, latent within this concept or practice of *kalatesa* is a tension, or the risk of *phit kalatesa* that questions or unbalances the authority of such social structure. The abandonment of manners is often attributed to Thai social problems, and the occupation and disruption of the city center by protesters is a very clear example of how that imbalance plays out (2000: 38-39). Thailand's cryptocolonial context necessitated the incorporation of western conceptions of space and politics into the city. Respecting *kalatesa* in the context of urban and political development in Bangkok then reflects not only knowledge of Thai conceptions of hierarchy, but also the selective incorporation of knowledge of the West.

Figure 2: Displays of Modernity in 1908 (Wisarut 2009)

In the mid- to late-nineteenth century, King Chulalongkorn (Rama V) and his father King Mongkut (Rama IV) made many policy decisions during their reigns that were influenced by Western expansionism, including the concession of trading rights to Western powers.[2] Consequently, the reigns of King Chulalongkorn and his father played a key part in the infiltration of Western architectural forms and social practices, physically inscribing the landscape of Bangkok by integrating Western influences into the preexisting political and urban system. The first Thai king to visit Europe in 1897, King Chulalongkorn developed expansion programs and composed an urban landscape with an emphasis on the West, building Dusit Palace in a neo-classical Italian Renaissance style, and eventually connecting it to the Grand Palace by Ratchadamnoen Avenue, a Champs Elysees–like northward expansion of the city.

Originally, the Siam Square area was developed near the Pathumwan Palace and Temple, which still rest on the north side of Rama I; this area served as an escape for the king from Rattanakosin and the Grand Palace. Today the name of the district, Pathumwan, remains defined by these key monumental spaces. Additional palaces, including Wang Sra Pathum, and Wang Phetchabun in 1916 and 1919 as well as landscaping that included a series of lotus ponds and a small hill, were eventually built in the area. In 1917 Chulalongkorn University was founded by King Vajiravudh (Rama VI), the son of King Chulalongkorn, on land just south of present-day Siam Square. Chulalongkorn University would eventually lease the land that is now known as Siam Square. Therefore, today one cannot ignore the underlying influence of these monarchical monuments on the Siam Square and Ratchaprasong Intersection.

The 1920s to 1930s was a period of revolution and military dictatorship; but while the installation of the constitutional monarchy presumably meant a major change in the balance of political and economic power within Thailand, few substantial physical changes to the social hierarchy actually occurred in the early years of the Phibun government, leaving the majority of the nobility's land holdings intact (1938-44).[3] The cultivation of desirable Thai etiquette and mannerisms was a key part of developing power. The government of Prime Minister Phibuhn Songkram (1938-1944 and 1948-1957) drew guidelines for appropriate behavior that emphasized high culture, which according to Penny Esterik was an explicit acknowledgement of the need to keep surface appearances and a sense of order amongst the chaos and disorder that raged after the 1932 coup (2000: 39). The Democracy Monument was built in 1939 to commemorate the 1932 revolution that established a constitutional monarchy, and was intended by the military ruler, Plaek Pibulsonggram, to symbolize the center of a Westernized Bangkok. His desire at the time was that the Thanon Ratchadamnoen and the Democracy Monument would become the Champs Elysees and Arc de Triomphe of a Westernized Bangkok. The symbolism embedded in the monument is both Buddhist and European in origin, but naturally neglects to include any references to the monarchy. The government even went so far as to hire westerners like the Italian Corrado Feroci to implement relief sculptures around the base of the Democracy Monument. The *naga* or protective snake that is found around many temples is a part of Hindu and Buddhist mythology is also a key part of the reliefs surrounding the monument. Once again, the military dictatorship, like the prior monarchy, harnessed Western monumental forms in building national symbols to reassert a Thai national identity within a wider global network of power. The Democracy Monument would come to serve as a centerpiece for political confrontation over the rest of the century.

Figure 3

From 1932 until into the 1980s and 1990s, with short periods of civilian rule, the military had a stronghold on political power. In 1937, National Stadium was built on the west side of the Rama I and Phayathai intersection (image by David Luekens), exhibiting an eclectic design that merges neoclassical columns with Thai symbols (as you can see, this includes imagery of the royalty). While Japan greatly influenced Thailand during World War II, the United States used Thailand as a major strategic base throughout the Vietnam War. This period was marked by urban expansion with the Litchfield Plan, which expanded the road and highway network in and around Bangkok, including the Siam Square area.[5] Siam Square rests squarely between the easternmost end of the old center of the city—which encompasses Rattanakosin Island, the

Democracy Monument, and the Victory Monument—and the western end of Sukhumvit Road. Running almost to the Cambodian border, this area was not developed until the post-World War II era, and it continues to serve as the backbone of modern Bangkok. In the 1980s the department store rose in popularity, and "investors developed a new concept of the shopping center, a place composed of department stores, shops, offices, hotels, movie theaters, theme parks, and exhibition and convention halls" (Kaewlai, 2007: 4). Today the shopping center is key to urban social activity and defines the physical program of Siam Square and the surrounding area.

The 1997 economic crisis brought foreign retailers who opened many big-box retail stores throughout the city, again echoing the relationship Thailand has had since the mid-nineteenth century with other foreign powers.[4] Consequently, the only shopping center in the Siam Square Ratchaprasong area that opened according to plan in 1997 was Siam Discovery, one mall amongst several of the Siam shopping malls owned by the Siam Piwat retail company of which the Princess Sirindhorn and King Bhumibol are major shareholders. These investments are in addition to the fact that the commercial interests of the monarchy are tied up in the income gained from these properties, given that they are located on top of Crown Property Beaureu land (Unaldi, 2014: 385). During this period, in 1998, Thaksin Shinawatra, a businessman who made his fortune as a telecommunications tycoon, founded the Thai Rak Thai (Thai Love Thai) party, which took power in 2001. His neo-liberal economic policies, known as Thaksinomics, further opened Thailand's economy to the global market, and gained the support of the rural poor through the first universal health care initiative in Thailand (the "30baht-per-visit" scheme). In encouraging the development of the retail sector in areas like Siam Square, Bangkok was positioned as a leading shopping hub, comparable to other capital cities in Asia (Kaewlai, 2007: 128). Within the past twenty years, the center of political and economic focus, and consequently the center of the city, has often centered on Siam Square rather than Sanam Luang, Ratchadamnoen, or the Democracy Monument (Sanam Luang is a large lawn in front of the Grand Palace where markets, demonstrations, and various public activities have occurred historically).

Mall as Monument, Mall as Everyday

With the emergence of a strong middle class at the end of the twentieth century, malls became dominant spaces within the city. Some of the ways in which the space of the mall interacts with the surrounding area and local practices, affirms its position within a history of monumental spaces built to emphasize the power of certain groups within the city. The monumentality and symbolic power of the mall is reinforced through selective accentuation and regulation, found in Herzfeld's concept of cryptocolonialism. In other words, attempts to legitimize these spaces within the urban landscape results in the banning and incorporation of particular local practices in order to keep *kalatesa* or more importantly *phit kalatesa* in check.

The blending of Western and Thai conceptions of politics and space forces one to reexamine the locality of this space as not simply a juxtaposition or opposition between global and local groups and spaces, but as an interweaving of monumental and everyday spaces with the mall. While the other monumental sites across the city embody the mixing of global and local forms as products of the state or powerful groups, specifically as monuments and parade spaces, the mall, as a space of consumption developed by corporate actors, is not entirely similar given its conscious relationship with low classed spaces like the street market.

In the following section, I use ethnographic research performed during the summer of 2011, in the Siam Square Ratchaprasong area and with migrant construction workers living on the peripheries of Bangkok, to capture the ways in which this area transgresses the boundaries distinguishing the global from the local and how certain tensions manifest in that transgression. Street markets, infrastructural development and spiritual spaces serve as moments of interruption and insertion, where monumental architecture and everyday urbanism interact. These spaces are connected to the mall through a cryptocolonial context where class tensions are enacted within the selective blending of global form and practices with and by the local context. While the spiritual and infrastructural spaces described continue the Thai tradition of the selective incorporation of high class Thai space with western form, the interweaving that occurs relative to the street market is distinct in that it is a low class space. Therefore the variety of ways in which hybridity is found within this portion of Bangkok's urban fabric makes it an aesthetically and politically distinct urban space. The space of the mall, which dominates this symbolic center of economic and political power for the country, facilitates this hybridity on a monumental and everyday scale.

Spiritual and Religious Space

Some of the most visually poignant places in and around Siam Square, visible from the Skywalk, are the shrines that accompany office buildings and homes alike. Such shrines are built alongside larger complexes like Siam Paragon, as well as many of the low-density shops in Siam Square. Shrines are, in fact, based on a traditional cluster house (series of rooms organized around a central terrace) or temple in form, within which dwells a spirit who protects the inhabitants of the main house. While the shrine is erected for the protection of buildings regardless of their function, it houses the spirit that exerts a power over the surrounding area. The shrine and the mall become intertwined, with the shrine as the protectorate or provider of luck and the office building or mall as insuring the continued relevance of such spaces in Thailand. While it may seem contradictory to find a shrine standing next to a Starbucks, these structures demonstrate the strong, continued presence of spiritual spaces within the new core of city life. The Erawan Shrine is one of the most important spiritual sites in the Siam Square area, and is an extremely popular place to pray for luck, especially for Chulalongkorn University students wishing for good luck on upcoming exams. The shrine was built in 1956, when the Erawan Hotel experienced delays in its construction. It was later demolished and replaced in 1987 and the shrine remains quite popular. This relationship between the shrine and the modern space of the Erawan Hotel and mall that would replace it is not far from of the way in which the military government incorporated religious symbolism like the *naga* into the Democracy monument. This particular hybrid relationship between the mall and shrine is one of protection. Within these hybrid spaces, the insertion of Western forms reasserts the importance of spirituality and religion, whilst the spiritual spaces and religious symbolism protect such monumental spaces of the nation from harm.

On the one hand, the mall and the shrine naturally coexist with one another, on the other hand, the interactions between the temple and the mall, as competing spaces in Thai society, begins to reveal some tension within their relationship. This tension exists both between these two institutions as well as in the ways in which they are perceived as inherently classed. Middle-class Bangkokians interviewed often describe the mall as a form of public space for themselves; however, migrant construction worker informants primarily described their local temple in their hometown as a place for social or public interaction. One worker told me that he "has never been to any public space" but he has been to *wat* (temple). For marriage ceremonies "everyone just goes to temple." However, when I asked another construction worker if he had ever gone to the temples that he defined as public space on his map of Bangkok, he pointed to his clothing and said that he would not be allowed to go to those temples because of the way he dresses. As the center of power in Thailand, the city's famous temples arguably stand more as monuments than public spaces. Therefore, although many of the construction workers I interviewed had lived in Bangkok for a number of years, they expressed a sense of exclusion from religious spaces in the city.

Wat Pathum Wanaram stands directly between two of the biggest malls in the area: Central World and Siam Paragon. Meditation temples like Wat Pathum Wanaram are quite popular, and it has become common for people to reserve a spot within the temple. The need to reserve spots and the ranking of temples as possessing a particular "status," exclusivity, or class is a contestable subject relative to Buddhist teachings of resisting materiality. Wat Pathum Wanaram is not considered a "normal" or "traditional" temple given its "modern" surroundings as well as its history as a part of the palace complexes that had founded the area.

Although it does not possess the same community one may find in a village temple, few doubt Wat Pathum Wanaram's continued presence in the area. The malls' neon glow that looms over the temple at night extends into its grounds during Buddhist holidays, with multicolored lights that illuminate the surrounding pathways. During such events, fences emblazoned with the Central World logo control the crowds that push through the temple grounds. During one visit, I sat down to talk with one of the volunteers working at Wat Pathum Wanaram, and he explained that Central World had offered to carry out preservation work on the temple. However, the offer by Central World was refused given that their conditions required stamping of the Central World logo onto the temple pillars. The donation of money and the repairing or refurbishing of temples is expected and usually involves families and individual members of a community. Similarly, names of these community members are often inscribed into the temple space; in fact, Buddhism encourages the addition and changes made to a site to gain merit, yet Central World was turned down when they sought to do so. Through improvements and the donation of chairs and fences, the mall is an agent through which the Central World group acts, but in attempting to inscribe their name in the temple to potentially gain capital and not necessarily solely merit, a line is drawn.

Although Bangkokian shoppers go to the temple or shrine alongside trips to the mall, many Thais are extremely conscious and at times wary of attempts to insert into or attach religious spaces to the mall. This debate over the insertion of temples into malls is directly connected with other corporate cultural practices that bring monks to provide talks or send their workers to weeklong meditation retreats, and also blur the divisions between corporate and religious spaces. This site, that preexists the insertion of the modern space of the shopping center, challenges the notion, most exemplified by Chinese megaprojects, that older layers of the urban fabric are always cleared to make way for global spaces like the mall. While it is clear that the mall and the temple are in tension with one another, it is through their coexistence and interdependences as spiritual spaces that such tension emerges.

Infrastructure and Form

Figure 4

The malls and infrastructure that emerge in the Siam Square area are a form of reordering space to fit a particular image of the "modern" city, as Ong describes in the case of the CCTV tower. The production of new or modern spaces is layered on top of the preexisting, with the luxury malls raised above street level, reasserting the vertical or hierarchical ordering of space mention earlier. This new infrastructure and form that dominates the Siam Square and Rachtaprasong area becomes a network between architectural and institutional monuments. Like the case of Wat Pathum Wanaram, these historical layers of infrastructure do not necessarily erase the preexisting layers of urban fabric and infrastructural systems. Infrastructure and form in this area are hybrid in that they layer upon previous systems and in their use continue to adhere to a hierarchy of high and low spaces. However, within these forms running through the space of the mall are tensions over competing groups attempting to dictate its movement as well as the lower class groups that may not have the same level of access to certain types of transportation.

Stepping off the Skytrain at the Siam Square station, one is hit by a wall of humidity and heat, forced to descend a set of escalators as a part of the massive wave of pedestrians. Although the ultimate destination may be the Bangkok Cultural Center that stands just across the intersection of Thanon Phaya Thai and Thanon Rama I. Because it is inaccessible via a crosswalk, one must take a winding path through the malls to get to the other side of the Skywalk. Siam Center and Siam Discovery are connected on the third floor, which directs the visitor via a winding maze of shops to the opposite side. Small signs, written in English, direct the pedestrian towards another mall or Bangkok Mass Transit System (BTS) station, each with their own little logo, offering little or no escape from this floating urban network of spaces dedicated to consumption. These walkways simultaneously limit the paths and experience of their occupants while also serving as a place where social status is reasserted.

Figure 5: (McGrath 2008, 77)

As in many cities, infrastructure is built and its pathway is decided through a balancing act between various institutional players manifesting themselves spatially within the city. The Skytrain and Skywalk connect these economic symbols of modernity, wealth, and luxury.[6] Figure six is a rendering of the paths of the Bangkok metro and Skytrain by Brian McGrath that reveals their strong give-take relationship with the retail development marked in red. The BTS Skytrain, which began operation in 1999, just after the Asian economic crisis, allowed retail operators to connect directly to the platform. This control over movement and the development of infrastructure by powerful groups in the city is visible in a number of projects including: King Mongkut's construction of Charoen Krung (Prosper the City) road outside the palace in 1862 as a response to a petition of European residents, and King Chulalongkorn's construction of over 120 new roads during his reign including Thanon Ratchadamnoen, originally built in 1899 to connect Dusit Gardens and its palace to the Grand Palace (both symbols of a progressive and civilized state).

Today, most people travel to the Siam Square area via car, bus, or Skytrain. While traveling to the Siam Square area via bus is an indication of poverty, those who arrive by car will drive directly into one of the enormous parking lots attached to the malls. The Skytrain is neither affordable nor an accessible distance from most poorer groups; however, they are able to visit the area via bus or a shared truck. So while both lower classes and higher classes may arrive at street level, those driving cars will still enter the site via the parking garage above the street level, reasserting once again the vertical spatial hierarchy at play. As discussed in the previous section, vertical movement is vital to the reiteration of social hierarchy in Thailand, and it manifests itself spatially in vernacular architecture like the temple and palace where zones and steps classify the social position and specific activities. The upper level and private areas stood for individuals of higher status including royalty, monks, and elders (Chitranukroh, 2006: 123).

While the primary means of transport into and out of the Siam Square area is by BTS or car, it was originally accessed by canal (see fifth figure). In early settlements the river was crucial for domestic, agricultural, and communication use; in fact, the first word of many town names is *bang* meaning "village settlement or group of shop houses built along a canal or river, or on the sea coast." Villages would expand linearly along the river, making administration difficult, but as a result of this development and growth of the village a temple would eventually be established as the community center (Chaichongrak et al., 2002: 20). To the north of Wat Pathum Wanaram is the San Saeb Canal, while to the south the Skytrain shoots above Thanon Rama I and Phaya Thai, and the Skywalk replaces the sidewalks. Therefore, the hierarchy of high and low spaces that the arrival of users of different classes seemingly adheres to is also reflected in the layering of old and new infrastructural spaces. The newer and more expensive means of arriving into the area are elevated above the older and cheaper systems.

This display of power and class through the spaces of movement in the Siam Square area extends to the Western architectonic forms within the malls themselves. The main entrance on Siam Paragon's southwest corner consists of a large plaza set between Siam Center and Paragon. Not surprisingly the design of the Siam Complex's primary pedestrian entry area is also raised above the street level. Rather than replacing the previous main entryway on the street level steps in front of Siam Center, upon the construction of the Skytrain the main entry was raised above the street level. The doorways of Siam Paragon are set into a curved glass wall that opens up to the large plaza. The architects of Siam Paragon have stated, "Our design delivers intelligent solutions and balances out the commercial, social and ecological demands. Our aspiration is to create an exciting retail experience, emphasizing pedestrian streets and plazas."[7] By harnessing air-conditioning in this exceptionally hot climate, a particular notion of public life as "pedestrian streets and plazas" can be reflected in the design; however, the iconic forms of the plaza or pedestrian street are not always used as intended, including the plaza outside the main entrance of Siam Paragon. During informal interviews multiple informants failed to include this plaza in their maps. This is not surprising given that pedestrian traffic flows directly across this open space. During the rainy season, crowds move through covered walkways between the two malls. Not only does the weather make this plaza an unattractive meeting space, but also the presence of surveillance and regulations as well as a lack of seating deters its use. Therefore, although the form of the mall itself consciously incorporates infrastructural spaces, this does not necessarily mean that they produce the public life imagined by the architects. The architects through their statements are entirely unconscious of the ways in which this space is actually used or how it creates another layer of urban fabric that is inevitably tied to what came prior physically and socially.

Politics and Aesthetics of Creativity

The Street Market

Figure 6

The street market, while a moveable or a space in transience, is without a doubt key to everyday life in Thailand. Markets range in size, content, and placement. The banning and/or regulation of such practices for the sake of *kwam civili* in Thailand demonstrates the selectivity that goes into the production of "formal" spaces. However, Siam Square and the protests that occur within it further complicate the distinction between the "informal" and "formal" when subversive forms of architecture and political groups take over this seemingly simple and structured area. The case of the street market clarifies that although within hybridity there is a certain tension, there is not always a juxtaposition between the seemingly distinct spaces at play. The relationship between the mall and the street market best exemplifies how the mall symbolizes the current sociopolitical present in Thailand. It is a moment in which the high and low class groups are forced to confront one another and not simply exist as present yet divided.

As the sun sets on Siam Square the *dalaat nat kon duurn* emerges, an "informal" street market that gets so crowded that movement through this area almost comes to a halt. During the day this sidewalk primarily consists of food stalls, or *raan ahaan kang thanon*, where fruit, meatballs, and coffee are often sold. Similarly, in the evening outside of Central World Plaza pops up a row of *som tum* restaurants, serving spicy seafood soup to those sitting at their makeshift plastic tables. Even "hiso" Thais have regular street vendors they frequent, making their drivers stop there to buy food. The night market run by the mafia is where the majority of shopping occurs in Siam Square. Nighttime is ideal for this space, given the much cooler temperatures. Vendors take advantage of the Skytrain above to store goods beneath its entryways, a safe place to keep them dry in case of rain during the summer months.

Street hawking is the most transient and contested form of the street market in the Siam Square and Ratchaprasong area, not unlike many markets found across the world. Although street markets are without a doubt a key space in everyday life in Thailand, found in urban and rural life alike, it is characterized as a "lower" social space, one that, at times, must be cleared or cleansed from the city. *Haperpeengloi*—meaning to put heavy things on your shoulder and travel and *peengloi* being a temporary kiosk—are temporary tables/kiosks that vendors move around with so, when the police come, they can wrap things up and leave very quickly. These are most often found on the skywalk but also occupy the entire city. Consequently, upon its banning from the city, *haperpeengloi* directly connects to the ways in which infrastructure dictates movement and reasserts power within the Siam Square area.

There is a long history of this type of banning in Thailand; the military leader Sarit Thanarat (Prime Minister from 1957-1963) also banned certain activities like pedi-cabs as "archaic and uncivilized" and attempted to erase informal practices such as street vendors. He concentrated on cleanliness and orderliness, striking down on crime, prostitution and drugs.[9] In addition, Alan Klima in his film *Ghosts and Numbers,* captures the efforts of the government in 2003 to ban lottery sellers, flower vendors, and poorer communities along the river in anticipation of an upcoming international economic summit. Like the restrictions on the use of space in the old city center in order to promote tourism, the banning of such practices would create a front for western tourists and officials. This is not far from Salwa Ismail's description of market vendors in Cairo where she describes the space of the market as in tension with other, more "legal" forms of entrepreneurialism. She argues that the practice of hawking represents the far-reaching intrusion of the security and police forces "into the details of everyday life" (Ismail, 2014: 273).

The relationship between street markets and powerful groups, however, is not always one of banning. There is a balancing that occurs in the constant clearing and reinstallation of such spaces within the city. During certain historical moments, monumental spaces were selected within the city to facilitate temporary markets. When discussing where people went for leisure thirty years ago, many pointed out that Sanam Luang was a place where you could fly kites or relax on the great expanse of lawn, arguing that the restrictions on street vendors in the area makes it a less attractive part of the city to visit. The weekend market known as Chatuchak or JJ market, located to the north of Bangkok, was initially located in Sanam Luang and founded by Field Marshal Plaek Phibunsongkhram in 1948, Prime Minister of Thailand (1938-44 and 1948-57). Not only have bans been set on the presence of market space in Sanam Luang, but also gates to prevent protests are now inserted around the park. These restrictions are primarily motivated by historic preservation projects in and around this old city center that are directed towards tourism.

Dalaat nat kon duurn, visible outside many malls in Thailand, benefit the malls by attracting crowds that are searching for deals who often enter the malls for food or entertainment. *Dtalaat nat kon durrn* is an example of the hybridity of this area, both physically and socially, that fluctuates between incorporation and rejection from "formal" structures of architecture and politics. This vibrant street scene is quite different from the stark hallways of Siam Paragon and Central World. Color and distinctive display is key to catch the eyes of passersby who slowly squeeze through the crowd to move from vendor to vendor. Daeng is one of a number of the vendors in this market who sells printed t-shirts. Sitting behind a rack of shirts as various passersby stopped and perused his selection, we discussed how he and his brother (who sells shirts at Chatuchak) started this business. Daeng began selling t-shirts at the market across the street from Siam Paragon three years ago, where he moved to a different location every day. While his brother sells shirts at Chatuchak—one of the largest the weekend markets in the country located on the northern edge of the city—he continues to sell these shirts in Siam Square as a branch of their business. Five months after he began selling in the area Daeng was assigned his current location from the mafia of Siam Square. By paying 6,000 baht per month for his spot, with a profit of 1,500 or 2,000 baht per night (50-67 dollars per night), the mafia then pays off the police to allow him, and other vendors, to stay there. Street markets are often categorized as a lower class phenomenon. However, in reality those running the lower density shops and selling their products in the market are not necessarily poor or from the rural northeast, they are often young Bangkokians who graduated from well-known universities—Daeng happened to study economics at one of the top universities in Bangkok. Daeng and his brother are, in fact, young entrepreneurs like many vendors in this particular market. They are busy promoting their own brand and making plans for the expansion of their business. Although supposedly outside both a "formal" architectural language and legal framework, the market remains both an essential part of urban life and shopping in this area. Those running the malls actually actively seek out the integration of these markets around the malls through the deals made between the mall owners, mafia and police. The street market is consequently planned into the mall through such practices; they are attractive in their ability to draw crowds in search for deals, and who then enter the malls for food or entertainment.

This incorporation of the mall and the street market is taken a step further in the case of Mah Boon Krong, or MBK, a shopping center that is distinct relative to the surrounding luxury malls in that it literally inserts a market space into an air-conditioned bubble. Many compare MBK to Platinum and several other shopping centers that are not directly connected to the Skytrain but stand just north of Central World and Siam Paragon. The shops in MBK spill into the walking area, creating a maze of sellers of similar goods, while Siam Paragon's shops create a very sterile experience for those meandering through its corridors. This is visible in the very plan of each floor where shops are condensed together with little hallway space. However, MBK is not entirely different from the surrounding luxury malls; other entertainment venues, including a movie theater and bowling alley are also set within its complex. As one ascends through MBK, the program shifts from fashion to more luxury products including leather and jewelry, to technology, and finally to the entertainment complexes of movie theater and bowling alley. While the actual circulation and floor plan creates a less "chaotic" market-like atmosphere, Siam Paragon and other malls in the area similarly organize

Politics and Aesthetics of Creativity

their program by floor and include various entertainment complexes. In fact, upon its opening, MBK was described by one of the major business magazine stated that MBK "'sheds new light on the further growth and expansion of Bangkok as a truly modern Asian city,'" demonstrating that its street-market atmosphere is not necessarily antithetical to the hybrid imaginary of the Thai nation-state (Wilson, 2004: 107).

In his book on Jakarta, Abdoumalique Simone describe an illegal produce market that stand juxtaposed but inevitably in dialogue with the official wet market at Pasar Tambora where "local authorities in the surrounding subdistricts have invested heavily in the market, bringing a "legal order" to its functioning in the absence of any official sanction from above" (Simone, 2014: 52). Whereas "legal order" is brought to these "illegal" operations in Jakarta through the local authorities, the mall owners themselves actively encourage the operation of such markets in Bangkok. For him this juxtaposition of the "informal" and "formal"/"legal" and "illegal" market space is a phenomenon of the "near-south." Simone describes how in Jakarta these night markets do not open until the "official" markets close; however, in Bangkok it is important that they share operating hours in order to draw crowds. While at times the juxtaposition of the mall and market in Siam Square is not far from the condition described by Simone, the interaction between the street-market and mall functions in a different manner that is actually intertwined with the success of the luxury malls. MBK and the night market across from Siam Paragon demonstrate that the street market vendors are not the only groups that "know how to stage their illegality" but also the mall developers recognize and want to capitalize on the incorporation of these everyday spaces (2014: 48). In other words, although the street market and Siam Paragon are distinct spaces, this does not necessarily mean that they are in competition or exist solely in their juxtaposition with one another.

Conclusion: The Explosion of Tension and Exaggeration of Spatial Divides

The ways in which the markets, spiritual spaces, and infrastructure interact with the space of the mall demonstrates how the monumentality and the everydayness of malls facilitate a multitude of ways in which hybridity manifests itself in this portion of the city. Moments in which the space of the market is shut down or malls are limited in the extent to which they engage with spiritual spaces are dependent upon whether they are perceived as tarnishing the image of Thailand as a "civilized" nation-state. In other words, hybrid spaces are considered acceptable according to those in power until the moment in which they are perceived as imbalanced or *phit kalatesa*. Therefore, the protests and the occupation of this area are not simply a consequence of the form as a monumental moment within the city, but an explosion of the preexisting class tensions that are embodied by Thai citizens and lay latent within this symbolic national space.

Political unrest often emerges around monuments created by various social groups that dictate a particular image of a nation-state or regime. Siam Square is an extremely dense and politically potent portion of the city; its monumentality symbolizes a much larger geopolitical network than Thailand alone. In this core of the city the old and new are not only merged but also layered, creating a labyrinthine space where seemingly distinct urban practices are interwoven. The lower class groups and practices that are part of the Mall as an everyday space, but kept hidden from the global image of Thailand, took over this "formal" space in 2010. Those who took to the street were commonly referred to as "stupid, ignorant, fussy babies," and some informants even argued that the military should have attacked the Siam Square area earlier and struck the protestors down. The roles and meaning of religious, infrastructural, and market spaces and practices in the Siam Square and Ratchaprasong area all became exaggerated in the context of the political conflict.

During the protests, the Red Shirts became the informal as unwanted and dirty space outside of the legal, which is normally hidden from view.[8] They took over the street and allegedly burned Central World mall. Subversive informal spaces do not just manifest in the form of the street market, but also characterize people and groups that, more often than not, blur and challenge the lines drawn around supposedly formalized spaces, both political and architectural. While there is a certain everydayness of newer or modern spaces in the Siam Square area, including the Skytrain and Skywalk that are key to the daily lives of many upper-middle-class Bangkokians, their monumentality was brought to the forefront during the protests as the malls were turned into fortresses. Upon its occupation, being a center for transportation and flow through the city both by car and public transportation, the rest of the city was frozen. However, many Bangkokians continued to work, flying via Skytrain over the barricades protesters set in the roadways. Consequently, the malls and their modern infrastructure became a fortress-like network that funnels workers into and out of the skyscrapers. While the malls remained open, for a significant portion of the protests the smaller shops on the street level were closed. Those who could make it past the barriers were the motorcycle taxi drivers and those riding above the barriers via Skytrain, two vastly distinct groups in Thai society. Therefore the infrastructural networks hierarchical and layered nature became exaggerated upon the occupation.

While the spiritual spaces of Bangkok are also considered classed according to my informants, in the context of the protests, their protective rather than classed characteristics came to the forefront during the areas occupation in 2010. After the military crackdown on the area, protestors ran into Wat Pathum Wanaram as a safe haven. One of the biggest scandals revolving around this military crackdown was the discovery of six bodies within the temple grounds. Thus, a debate ensued between different political parties over whether those individuals were killed within the temple grounds by the military. Religious spaces are characterized as zones within the urban fabric that are supposed to be free of or able to control agents of political and economic will, whether that be the mall or a political group. While Wat Pathum Wanaram is supposedly not a "traditional" temple and a space through which a debate about the changing role of religion in Thai society is discussed on a daily basis, its continued presence as a protective agent and independence from the encroaching "modern" space of the mall is reasserted when it becomes a safe haven for protestors. Similarly, in March 2006 a man attacked the Erawan Shrine with a hammer and was consequently beaten to death by a crowd in the street. Upon the reopening of the Erawan Shrine, Sondhi Limthongkul, a political opponent of Thaksin, accused Thaksin of being involved with the vandalizing of the statue. One blog stated: "Sondhi, a media-tycoon who was then leader of the People's Alliance for Democracy (the Yellow-Shirts)—who himself once led a water-sprinkling ritual at the Government House with intent to exorcise ill-intended spirits—claimed the vandalizing event may have been a evil plot designed by Thaksin so he could then replace the statue in order to somehow gain black magic powers associated with the Erawan Shrine" (Sullivan, 2010).

This piece opens up the question of whether the protests and consequent deaths on May 19, 2010, were a result of a curse set on the shrine by the deranged man. The blogger continues to state that the depiction of Brahma in the shrine "witnessed the unfolding of the entire event," including the burning of Central World and the deaths of over eighty people. Concluding, the post expresses hope that if there was any curse in connection with the destruction of the shrine, it has "Fulfilled its evil intent and no longer holds the citizens of Bangkok hostage to further doom. Certainly the ongoing, deeply hateful division of modern Thai society alone has served as retribution enough. Solutions now lie not in anything related to powers of the 'super-natural' (or perhaps political) world; but in color-blind, human reconciliation of some degree" (Sullivan, 2010). The protests cause individuals to look back at these potent social spaces as an explanation for the surrounding violence. While on the one hand the shrine protects these modern spaces, acting upon the modern space of the mall, ill-intentioned political agents also have the ability to act upon it.

In this chapter I examined how distinct moments in the Siam Square area facilitated by the space of the mall reflect a bridging between local and global, everyday and monumental, and informal and formal forms of architecture and urbanism. The street markets, infrastructural, and spiritual spaces that perform this bridging and blurring are saturated with tension that exploded in the context of the protests. The mall and its surroundings are not a simple matter of "Junkspace," as Rem Koolhaas would call it, but part of an urban system that is connected to a unique history of protest, political hybridity, and urban development.

This portion of the city continues to be a space of political action; Thais took to the streets in protest of the Amnesty Bill that Thaksin Shinawatra's sister Yingluck Shinawatra, the now ex-Prime Minister of Thailand, attempted to pass. This bill would have dropped the charges not only against Thaksin Shinawatra but also against those accused of murdering the ninety-one protesters who died in 2010. This bill upset Yellow Shirt supporters as well as Red Shirts, who also took to the streets of Bangkok to protest the Amnesty Bill. In addition, recent protesters have harnessed pop-culture in acts of dissent against the current military regime. Each of these protests are not at all of the same demographic that took over the Siam Square area in 2010. The continued use of this space as one in which dissent is expressed only further confirms the political significance of this area and only time will tell the new and diverse ways in which this global space will continue to serve as a stage on which these tensions play out.

Notes

1. Foucault, 1975; Rabinow, 1989.
2. In 1855 King Mongkut (Rama IV) signed The Bowring Treaty with Britain as well as many similar treaties with other Western powers that "reduced import-export duties, abolished customary shipping fees and removed many trading monopolies still in place on export products" (Askew, 2002: 27).
3. Askew explicates, "Apart from changes within the top echelons of the power elite and their enhanced access to material and status rewards, there was no redistribution of wealth—no fundamental change to the socio-economic structure of the countryside or the city flowed from the revolution" (2002: 47).
4. The philosophy of sufficiency economy was proposed by King Bhumibol Adulyadej during the 1997 Asian economic crisis.
5. By 1960 large canals had been filled in for the construction of new roads, and the Litchfield Plan, developed in the 1960s and 1970s by American planners as an automobile-centric plan for the expansion of Bangkok, continued this trend. Its emphasis on infrastructure like highways created a gridlock system of traffic with little thought for the development of a smooth, scalar shift from the soi to highway system. The Litchfield Plan is one of many examples of the strong influence of the United States in Southeast Asia and in particular Thailand during the Vietnam War.
6. Wisarut, "Forum," *2Bangkok.com News*, 16 September 2009, http://2bangkok.com/forum/showthread.php?3777-Bangkok-Photo-in-1967-by-Khun-jarcje/page4 (accessed 20 March 2012).
7. Rosalynn Poh, *City & Country: J + H Boiffils to Design Iconic Development in Bukit Jelutong*, 6 December 2010, http://203.115.229.228/edgemyjoomla/property/179101-citycountry-j-h-boiffils-to-design-

iconic-development-in-bukit-jelutong.html (accessed 20 March 2012).

8. The protesters reassert their agency by taking over the modern core of the city, an icon of wealth and consumerism from which they do not reap the "benefits." They are the counterpublic as described by Warner: "A counterpublic maintains at some level, conscious or not, an awareness of its subordinate status" (2002: 119).

9. In her book *Purity and Danger*, Mary Douglas demonstrates how distinct, culturally driven structures influence and define what we know and understand to be dirt, making it "matter out of place." To clean is a positive reordering of our environment; Douglas states, "In chasing dirt, in papering, decorating, tidying we are not governed by anxiety to escape disease, but are positively re-ordering our environment, making it conform to an idea. There is nothing fearful or unreasoning in our dirt-avoidance: it is a creative movement, an attempt to relate form to function, to make unity of experience" (1966: 2). Groups from a particular position of cultural knowledge or power perform this conscious "creative movement" for Douglas. This argument is at times applicable to the development of architectural and urban projects, often of monumental scale, where they are not necessarily designed with a focus on the locality but on norms emergent from Western aesthetic sensibility.

Bibliography

Althusser, Louis. *Ideology and Ideological State Apparatuses.* London: New Left Books, 1971.

Architecture and Design. www.boiffils.com (accessed 20 March 2012).

Architecture and Urban Facilities. 23 November 2004. http://www.skyscrapercity.com/showthread.php?t=153491 (accessed 20 March 2012).

Askew, Marc. *Bangkok: Place, Practice and Representation.* London and New York: Routledge, 2002.

Barthes, Roland. *Empire of Signs.* Translated by Richard Howard. New York: Hill and Wang, 1982.

Becker, Lynn. *An Interview With Rem Koolhaas.* http://www.lynnbecker.com/repeat/OedipusRem/koolhaasint.htm (accessed 20 March 2012).

Boddy, Trevor. "Underground and Overhead." In *Variations on a Theme Park: The New American City and the End of Public Space*, edited by Michael Sorkin. New York: Hill and Wang, 1992.

Boontharm, Davisi. *Bangkok: Formes du commerce et evolution urbaine.* Paris: Recherches, 2005.

Buchli, Victor, and Gavin Lucas. *Archaeologies of the Contemporary Past.* London and New York: Routledge, 2001.

Caldeira, Teresa P.R. *City of Walls: Crime, Segregation, and Citizenship in Sao Paulo.* Berkeley: University of California Press, 2000.

Calhoun, Craig, ed. *Habermas and the Public Sphere (Studies in Contemporary German Social Thought).* Cambridge, MA: MIT Press, 1992.

Chaichongrak, Ruethai, Somchai Nil-athi, Ornsiri Panin, and Saowalak Posayanonda. *The Thai House: History and Evolution.* Trumbull, CT: Weatherhill, Inc., 2002.

Chitranukroh, Jayanin, and Vorasun Buranakarn. "Sentiment in Traditional Thai Architecture." *Nakhara: Journal of Oriental Design and Planning* (2006): 117-32.

Crawford, Margaret, and Setha Low. *Variations on a Theme Park: The New American City and the End of Public Space.* Edited by Michael Sorkin. New York: The Noonday Press, 1992.

Crawford, Margaret. *Everyday Urbanism: Margaret Crawford vs. Michael Speaks* in Michigan Debates on Urbanism volume 1, edited by Rahul Mehrotra. New York: Distributed Arts Press, 2005.

Dohring, Karl. *Buddhist Temples of Thailand.* Bangkok: White Lotus, 2000.

Douglas, Mary. *Purity and Danger: An Analysis of the Concepts of Pollution and Taboo.* New York: Routledge, 1966.

Evers, Hans-Dieter, and Rudiger Korff. *Southeast Asian Urbanism: The Meaning and Power of Social Space.* New York: St. Martin's Press, 2000.

Fencing Off Sanam Luang. 1 August 2011. http://2bangkok.com/fencing-off-sanam-luang.html (accessed 20 March 2012).

Floor Plan MBK Center. http://www.mbk-center.co.th/en/floorplan/ (accessed 20 March 2012).

Foucault, Michel. *Discipline and Punish.* New York: Random House, 1975.

Frampton, Kenneth. "Reflections on the Autonomy of Architecture: A Critique of Contemporary Production." In *Out of Site: A Social Criticism of Architecture,* edited by Diane Ghirardo. Seattle: Bay Press, 1991.

— "Towards a Critical Regionalism: Six Points for an Architecture of Resistance." In *The Anti-Aesthetic: Essays on Postmodern Culture,* edited by Hal Foster. New York: New Press, 1998.

Gluckman, Ron. *Bangkok's hip mini-malls.* 2007. http://www.gluckman.com/BangkokShoppingMalls.html (accessed 20 March 2012).

Habermas, Jurgen. *The Structural Transformation of the Public Sphere: An Inquiry into a Category of Bourgeois Society (Studies in Contemporary German Social Thought).* Cambridge, MA: MIT Press, 1999.

Hack, Gary. "Place-Making and the New Mobility of Asian Cities: The Bangkok Plan." In *Public Places in Asia Pacific Cities: Current Issues and Strategies*, edited by Pu Miao. Dordrecht, Boston, London: Klower Academic Publishers, 2001.

Harvey, David. *Consciousness and the Urban Experience*. Baltimore: Johns Hopkins University Press, 1985.

Hernandez, Felipe, Peter Kellett, and Lea K. Allen, eds. *Rethinking the Informal City: Critical Perspectives from Latin America*. Remapping Cultural History 11. New York and Oxford: Berghahn Books, 2010.

Herzfeld, Michael. "The Absent Presence: Discourses of Crypto-Colonialism." *South Atlantic Quarterly* 101, no. 4 (2002): 899-926.

— "Paradoxes of Order in Thai Community Politics." In *Radical Egalitarianism: Local Realities, Global Relations*, edited by Felicity Aulino, Miriam Goheen, and Stanley J. Tambiah. New York: Fordham University Press, 2012.

Holston, James. *The Modernist City: An Anthropological Critique of Brasilia*. Chicago: University of Chicago Press, 1989.

Hoskin, John. *Bangkok by Design: Architectural Diversity in the City of Angels*. Bangkok: Post Books, 1995.

Ismail, Salwa. *The politics of the urban everyday in Cairo: Infrastructures of oppositional action*. In The Routledge Handbook on Cities of the Global South, 2014. Susan Parnell and Sophie Oldfield, editors. London: Routledge. Pp. 269-280.

Kaewlai, Peeradorn. *Modern Trade and Urbanism: The Reciprocity between Retail Business and Urban Form in Bangkok and its Periphery*. Cambridge, MA: Harvard University, 2007.

Kanna, Ahmed. *Dubai: The City as Corporation*. Minneapolis: University of Minnesota Press, 2011.

Koolhaas, Rem. "Junkspace." *Obsolescence* 100 (October 2002): 175-90.

— "Whatever Happened to Urbanism?" *Design Quarterly*, no. 164 (1995): 28-31.

Kurlantzick, Joshua. *The Dangerous Use of the Thai Monarchy.* 1 December 2010. http://blogs.cfr.org/asia/files/2010/12/20101201_ThailandKingProtes.jpg (accessed 20 March 2012).

Kusno, Abidin. *Behind the Postcolonial: Architecture, Urban Space and Political Cultures in Indonesia.* London and New York: Routledge, 2000.

Lin, Maya. *Boundaries.* New York: Simon & Schuster, 2000.

Lopes, Sal, and Michael Norman. *The Wall: Images and Offerings from the Vietnam Veterans Memorial.* New York: Collins Publishers, 1987.

Low, Setha M. "The Anthropology of Cities: Imagining and Theorizing the City." *Annual Review of Anthropology* 25 (1996): 383-409.

— "Indigenous Architecture and the Spanish American Plaza in Mesoamerica and the Caribbean." *American Anthropologist* 97, no. 4 (1995): 748-62.

Low, Setha, and Neil Smith. *The Politics of Public Space.* Edited by Setha Low and Neil Smith. New York: Routledge, 2006.

Luekens, David. 21 October 2012. "Bangkok's National Stadium." http://www.travelfish.org/blogs/thailand/2012/10/21/bangkoks-national-stadium/ (accessed 19 October 2013).

Marshall, Richard. *Emerging Urbanity: Global Urban Projects in the Asia Pacific Rim.* London and New York: Spon Press, 2003.

McGrath, Brian. *Digital Modelling: For Urban Design.* West Sussex: John Wiley & Sons Ltd., 2008.

Mitchell, W.J.T. *Art and the Public Sphere.* Chicago: University of Chicago Press, 1992.

Mydans, Seth. "Government of Thailand is Promised Support." *New York Times.* 18 April 2010. http://www.nytimes.com/2010/04/19/world/asia/19thai.html (accessed 20 March 2012).

The Nation. Politics Chronology. May 21, 2010. http://www.nationmultimedia.com/admin/specials/sound/file/36-site-attacked.jpg (accessed 20 March 2012).

Oxford Dictionary. http://oxforddictionaries.com/definition/liminal?region=us&q=liminality#liminal__3 (accessed 20 March 2012).

Ong, Aihwa. "Hyperbuilding: spectacle, speculation, and the hyperspace of sovereignty." In Worlding Cities – Asian Experiments and the Art of Being Global, edited by Ananya Roy and Aihwa Ong. Malden: Wiley-Blackwell, 2011.

Parr, Adrian. *Deleuze and Memorial Culture: Desire, Singular Memory and the Politics of Trauma.* Edinburgh: Edinburgh University Press, 2008.

Peck, Grant. "Thai King Speaks for First Time During Crisis: Ailing 82-year-old monarch fails to address chaos some call class warfare." *MSNBC.* 26 April 2010. http://www.msnbc.msn.com/id/36784752/ns/world_news-asia_pacific/t/thai-king-speaks-first-time-during-crisis/#.T4rcUOlWpGA (accessed 20 March 2012).

Purnell, Newley. "Mixed moods: Tourists party on Khao San Rd near shrines for Bangkok protest dead." *CNN GO.* 12 April 2010. http://www.cnngo.com/bangkok/none/despite-deadly-clashes-tourists-party-bangkoks-khao-san-road-876182 (accessed 20 March 2012).

Rabinow, Paul. *French Modern.* Cambridge, MA: MIT Press, 1989.

Sassen, Saskia. "The Global Street Comes to Wall Street." *Possible Futures.* 22 November 2011. http://www.possible-futures.org/2011/11/22/the-global-street-comes-to-wall-street/ (accessed 20 March 2012).

Scott, Rachelle M. *Nirvana for Sale?: Buddhism, Wealth, and the Dhammakaya Temple in Contemporary Thailand.* Albany: State University of New York, 2009.

Siam Paragon Development Co. Directory. 2010. http://www.siamparagon.co.th/v9/directory.php (accessed 20 March 2012).

Simone, AbdouMaliq. *Jakarta – Drawing the City Near.* Minneapolis: University of Minnesota Press, 2014.

Smith, Neil. "New City, New Frontier." In *Variations on a Theme Park: The New American City and the End of Public Space,* edited by Michael Sorkin. New York: Noonday Press, 1992.

— *The New Urban Frontier: Gentrification and the Revanchist City.* New York: Routledge, 1996.

— *Uneven Development: Nature, Capital and the Production of Space.* Athens: University of Georgia Press, 2008.

Smithsonian Institution. *Reflections on the Wall: The Vietnam Veterans Memorial.* Harrisburg, PA: Stackpole Books, 1987.

"Spiegel Interview with Dutch Architect Rem Koolhaas." 13 August 2006. http://www.spiegel.de/international/spiegel/0,1518,408748,00.html (accessed 20 March 2012).

"Sra Pathum Palace - the palace museum." Thread by Mel on The Paknam Web Forums. February 2006. http://www.thailandqa.com/forum/showthread.php?25552-Sra-Pathum-Palace-the-palace-museum (accessed 20 March 2012).

Sthapitanonda, Nithi, and Brian Mertens. *Architecture of Thailand: A Guide to Traditional and Contemporary Forms.* London: Thames and Hudson, 2006.

Sullivan, James. "The Erawan Shrine Curse and the Red Shirt Demonstrators." *Thailand Law Forum.* 4 June 2010. http://www.thailawforum.com/erawan-shrine-curse-2.html (accessed 20 March 2012).

Tambiah, Stanley J. *World Conqueror and World Renouncer: A Study of Buddhism and Polity in Thailand Against a Historical Background.* Cambridge: Cambridge University Press, 1976.

Turner, Victor. "Frame, Flow and Reflection: Ritual and Drama as Public Liminality." *Japanese Journal of Religious Studies* 6, no. 4 (December 1979): 465-99.

Unaldi, Serhat. "Working Towards the Monarchy and its Discontents: Anti-royal Graffiti in Downtown Bangkok." *Journal of Contemporary Asia*, 44, no. 3(2014): 377-403.

Van Esterik, Penny. Materializing Thailand. Oxford and New York: Berg, 2000.

Venturi, Robert. *Complexity and Contradiction in Architecture.* New York: Museum of Modern Art, 1977.

Walters, Steve. "Siam Paragon and Siam Ocean World." *Thailand Musings.* 20 January 2008. http://thailandmusings.thaivisa.com/siam-paragon-and-siam-ocean-world/ (accessed 20 March 2012).

Warner, Michael. *Publics and Counterpublics*. Brooklyn: Zone Books, 2002.

Widodo, Johannes, ed. *The Boat and the City: Chinese Diaspora and the Architecture of Southeast Asian Coastal Cities*. Singapore: Marshall Cavendish Academic, 2004.

Wilson, Ara. *The Intimate Economies of Bangkok: Tomboys, Tycoons, and Avon Ladies in the Global City*. Berkeley and Los Angeles: University of California Press, 2004.

Winichakul, Thongchai. *Siam Mapped: A History of the Geo-Body of a Nation*. Honolulu: University of Hawaii Press, 1994.

Wyatt, David K. *Thailand: A Short History*. Chiang Mai: Silkworm Books, 2004.

CHAPTER TWO:

Small-scale and Bottom-up: Innovating Creativity in Shanghai's City Center Quartiers
Ying Zhou

Introduction

In the decade since Shanghai became the "dragon's head" for China's economic liberalization, the former "Paris of the East" has moved astoundingly fast in its re-globalization. By the mid-2000s, the aspirations for becoming a creative city as part of its structural transformation, coupled with the recognition of its historic urban fabric, produced a compelling case for a reassertion of the city's entrepreneurial edge. From the initial bottom-up developments, like the conversion of neighborhood production sites in Tianzifang and former textile warehouses on Moganshan Road into artists' studios, the reuse of former manufacturing buildings developed into the much promoted "creative industries clusters" occurred just in time for the opening of Expo 2010, whose post-event site is touted to become the largest gathering of creative clusters in the world. The convergence of heritage protection through reuse of historic architecture and the programmatic upgrade into clusters for creative incubation offered a new mode of development for former manufacturing buildings on state-allocated industrial land, with the state playing an active role in their promotion and production. State designation of creative clusters not only formalizes global interest in the development of the creative city, but also localizes their production in the post-socialist urbanscape of state-owned enterprise (SOE) reorganization.[1]

Although the adaptive state has usurped pioneer examples set by bottom-up creative entrepreneurial efforts and institutionalized the processes into business plans for creative industry incubation, self-organized efforts nevertheless continue to innovate urban spatial production for cultivating the creative industries through the cosmopolitan agents who bring the latest "know-how" from transnational linkages. State-allocated land as a vestige of the planned economy has initially compelled creativity in their developments. But spatially, creative productions go far beyond the designated areas that are given a plaque and the formal endorsement by the municipalities and the districts.[2] The inner city neighborhoods as "urban breeding ground" (Christiaanse, 2010) for new economies affirm the urban qualities and spatial embedment for the complex value chains that constitute creative cultivation.

Existing studies on "creative industries clusters" have thus far focused on the officially designated areas in Shanghai, analyzing their role in economic aspiration (Wu, 2004; Keane, 2009), constitution of their creative content (Kong, 2009; Zhong, 2012), and their development processes (Wang, 2009; Zhong, 2009) as an indication for the local state's entrepreneurial role (Zheng, 2010 and 2011). But little is known about the organic developments by creative entrepreneurs in the modern neighborhoods at the periphery of the former Concessions that quickly became positionally central in the metropolitan area in the decade of expedited urban expansion to meet the pent-up spatial demand underserved since 1949. In lane-house neighborhoods like the Jing'an Villas and in the Anfu-Wuyuan area, cosmopolitan small entrepreneurs not only resist and evade institutional appropriation, exploit the vestiges of the planned economy, and innovate spatial reuse, but also seem to develop mechanisms for heritage protection and creative innovation.

Existing research of these urban neighborhoods introduces the larger context of recent transformation (Wan, 2009) and related historical legacy to image-making (Li, 2012),[3] or specifically looks at the effects of the formal redevelopments involving demolition-displacement and reconstruction (Ren, 2008; Yang and Chang, 2007); however, few have related the urban qualities of these legacy areas with their contemporary transformation into the crucial centralities for transnational talents with globalized creative knowledge in the metropolis. Adaptive reuse of the architecture typologies is increasingly being studied for the potential of these important centralities. In its adaptability, the modern urban structure of the neighborhoods—with open networks, semi-permeable block hierarchy, fine-grained cadastre as well as modern residential buildings—has been able to create a conducive habitat for new economies, with the necessary urban value chain of living, working, and encounters that are both local and global.

Creative reuses of residential architecture largely built in the 1930s are transforming these inner city neighborhoods into global-trend quartiers resembling the likes of New York's Williamsburg or Berlin's Prenzlauerberg, areas in the West known as being the harbingers of the creative class. Next to the old Shanghainese ladies gathered in gossip while tipping off the tails of bean sprouts in the lanes connecting the typical *lilong* housing,[4] young freelancers "tweet"[5] their way into their new co-working studios in a *lilong* house equipped with airbooks and Illy coffees. Across from the Uyghur restaurant, inserted into the terrace of another restaurant opened by inland-migrants promising authentic halal beef noodles, is the newly renovated boutique with a smartly chosen French and Chinese name. Located on the ground floor of a garden house, the boutique is serving a bubbly apéro as it showcases the international collection from a range of Danish-Chinese, UK-Taiwanese, French-Singaporean duos. Despite the international vibe that is ascribed to these neighborhoods, the procedural "informalities" of their spatial productions that confound Western presumptions of property rights, the institutional stability, and the clarity represented by the outward appearance of a globalizing and modernizing physical environment compel an interrogation of the agents, drivers, and the local frameworks for its urban spatial production.

From extensive site documentation, mapping analyses and semi-structured in-depth interviews conducted in 2012 and 2013, this piece offers three vignettes that contest the prevailing narrative of the state-directed creative city construction via designated cluster formation. In an urban production system where there is a top-down presumption that the production of spatial supply would generate anticipated demand—and this presumption is specifically projected on the making of clusters to give rise to the creative economies—a parallel creative ecology that is fluid and flexible offers a counterpoint of small-scaled, bottom-up entrepreneurs who strategically eschew the official "creative industries cluster" as an organizational model for systematizing creative labor, and who innovate production of spaces.

The first vignette about the designer-produced cluster of Anken Green, although officially inducted into the municipal list of "creative industries clusters," introduces the prevalent profit-driven development mode for creative cluster formation by offering a departure from its conventions. As a case for re-configuring the business model of a creative incubator, the concerns and design resolutions reveal the demand-side spatial production of creative space by and for small cosmopolitan creative entrepreneurs. In the second vignette of Jing'an Villas, the recently shut-down *lilong* house neighborhood that was at one point expected to become the next Tianzifang, the harboring of a creative ecology is more than just a productive agglomeration of synergies and knowledge transfers that harness creativity for revenue. The deliberate embedment of creative production into the residential fabric with amenities and conveniences evaded official appropriation; but, in resisting being churned into part of the urban growth machine, the creative entrepreneurs were in the end evicted. And finally, in the vignette of the Wuyuan-Anfu area, the importance of the urban neighborhood as social milieu nurturing the diversity of creative productions is shown in the small entrepreneurs' innovating spatial reuse, exploiting the vestiges of planned economy, and rejuvenating the heritage architecture that is falling apart. From building to block to neighborhood (Fig. 1), the innovations necessitated by constraints are revealed by examining the interactions of creatives as developers, entrepreneurs, residents, and users with different institutional players controlling space provision through tacit consent and existing conditions in the context of urban morphology and ownership tenure.

Figure 1

Anken Green: Alternative Business Plan for Creative Incubation

The architects Alex Chu and David San Roman started Anken Green in 2006 as an extension of their design firm, Enclave. Alex is an Australian of Chinese ancestry from Melbourne, and her partner, David San Roman, is Spanish with extensive work experience in the United States. Alex moved to Shanghai from Hong Kong in the early 2000s, and prior to that worked in Los Angeles with the international landscape firm EDAW, Inc. David had extensive experience with the New Urbanists,[6] having led the international design firm HOK's New Urban Studio. Their combined international experiences bring important insights to the spatial needs of international creative offices in Shanghai.[7]

As a design firm, Enclave began with locations in the vicinity of the Suzhou Creek, where many industries have been located since the early 1900s. The former industrial buildings—with high ceilings, big windows, floor throughspaces—have the type of ambience that creative firms prefer. And following the restructuring of state-owned enterprises (SOEs) in the mid-1990s, the buildings also offer low rents that are affordable to young design firms who often share the large spaces, enabling the collaboration that economists call horizontal and vertical networking. While few developers were interested in the empty warehouses at the periphery of the central districts, the crash of the dot-com boom in the early 2000s also meant there was little cash available for possible developments. Additionally, the SARS outbreak in 2003 slowed demolition and redevelopment projects. At the same time, in the mid-2000s, a growing number of international designers came to China in response to news of its developing market, many landing in Shanghai. They were undersupplied by the limited amount of work spaces for creative needs. Working from two other industrial spaces in the Jing'an district in 2004 and 2006, Alex and David were drawn to the Huai'an Lu industrial building, a disused warehouse owned by a local SOE at the edge of the district close to the Suzhou Creek. Having found a space for their own growing design office, the two designers undertook redevelopment of the entire building.

The development of the Anken Warehouse building (Fig. 2) was timely. The shift of Shanghai's city center from manufacturing to service industries, part of the fundamental restructuring of the metropolis, also coincided with the growing discourse on the competitiveness of global cities. The "rise of the creative class," as promoted by Richard Florida in 2002, as both indicator and instigator for a post-industrial, value-added economy required the accompanying spatial provision for their accommodation. In the mid-2000s, the establishment of the semi-governmental organization the Shanghai Creative Industries Center, at the behest of the Municipal Economic Commission, promoted a mode of development for the properties on state-allocated land—a unique form of ownership tenure that is a vestige of China's planned economy, and thus restricted in their development on the market—that supplied space to programmatic demand of the creative industries.

Small-scale and Bottom-up

Figure 2

As carefully dissected by the pioneering scholars of China's land marketization, the distinction between leasehold land and state allocated land (Yeh and Wu, 1996; Zhu, 2002; Hsing, 2006) establishes a dual land market with discrepancies in procedure as well as in price, creating an inherent inefficiency for urban development, in function as well as in physical structure. The 1990s saw the overzealous transfer of allocated land to leasehold land, accompanied by rampant unsanctioned change of use to profit-making commercial and residential functions with little benefit to the local district, not to mention the residents (Zhu, 2002); subsequently, allocated land, most of which is provisioned for industrial use, has been changed to leasehold land and restricted from ownership tenure as well as from land use change since the early 2000s.[8] The perpetual use rights of the structures on the allocated land may be held by SOEs, often expired industries that have shed their labor force but retained their management structure, and who thus could gain compensation from the district or municipal government should the allocated land be formally acquired. The SOEs are prohibited from redeveloping the buildings via demolition-reconstruction but have resorted to the leasing out of their buildings for a profit. For the dilapidated industrial structures—some historic architecture given heritage status in the mid-2000s, but largely ordinary buildings built after 1949—restrictions on official use change also ruled out their functional upgrade. Many of the structures that remain in the possession of the SOEs that no longer operate on their premises have little prospect for formal redevelopment or reuse.

However, with the 2005 "Eleventh Five-Year Development Plan of Creative Industries in Shanghai," the specific permission for use change was granted providing the functional upgrade would be made for creative industries, and on the condition that the official land use designation of industrial remains on the ownership certificate (Tang, 2013). This course of development for industrial allocated land held by SOEs offered a viable business plan for the informal redevelopment of the industrial structures while avoiding the transfer of land use rights and consequent fees that a formal development would have required. The SOEs retain their land use rights but can lease out their built structures for more lucrative, non-industrial functions under the guise of making available spaces in the city center for the creative industries.[9] At the same time, this satisfies the stipulation by central government that development avoid the over-commercialization of subsidized state assets, in the form of state-allocated land.

Between 2005 and 2006 the number of officially designated "Creative Industries Clusters" in Shanghai grew from the initial eighteen to seventy-five. The first batch included Tianzifang, a *lilong* house area with neighborhood industries that had already began transformation in the mid-1990s; and the Moganshan Lu area, former industrial buildings of the defunct textile industries that became art studios, also transformed and occupied by artists since the late-1990s. But the prototype of the business model that would be replicated in the ensuing developments was the Bridge 8 project, on a site owned by the Shanghai Automobile Group and developed by one of the former directors of the Xintiandi development. With a short renovation time and leases to creative offices—at rates well above lease rates from the tenure-holding SOE on industrial allocated land, but still slightly below office rental on the market—the investment return could be recouped, at full occupancy, within two to three years. Compared to the one-off return recouping in residential development, the differential between rising high rental profit and low leasing fee, locked in for the term with the SOE, generates continuous profit. This business formula for upgrading vacant former industrial areas into revenue-generating service industries is supplemented by financial and non-financial benefits offered by the competing districts, each eager to jump on the bandwagon of incubating creativity. The district, in return, also garners tax revenue from the commercial enterprises attracted to the site.[10] The official designation by the Municipal Economic Commission as a ceremonial prestige also brings an assortment of incentives, from tax holidays by the district to rental subsidies and expedited business registration, all attracting creative industries to settle in the designated areas.

Despite the seemingly simple business formula, the proliferation of creative clusters also saw a parallel growth of high vacancy rates. The maximization of profit in a short time-span meant the developers would set the rents too high to accommodate the creative enterprises their developments are intended to attract. Unsuccessful models are visible in places like High Street Loft, developed by the Sanqiang part of the ShangTex Group, where high turnover due to high rents and inadequate amenities has left the development with the feeling of dereliction. Even in the showcase project of the 1933 Old Millfun, which was touted as an industrial monument in 2005 and since then houses the Shanghai Creative Industries Center itself,[11] the higher than market price rentals and the spatial structure of the former slaughterhouse destined it to high turnovers and a steady vacancy rate.

Unlike most state-backed developments capitalizing on district incentives of developing creative industries clusters, Anken, being founded by designers for themselves, invested in the renovation by calculating backwards from what design offices such as their own can afford, making the development feasible from the outset.[12] This implicit understanding of the tastes as well as budgets of design offices subscribes not to the dominant and top-down logic of supply creating demand, but follows real demand rather than projected demand, aimed at maximizing revenue for both the developer and the district. With desks renting at 2,000 to 2,500 RMB per month and facilities for office infrastructure including meeting rooms and event spaces, which also serve as areas for networking with other creatives, the reasonable price for a start-up freelancer brought many initial tenants. Office rental rates of 4.5 to 5 RMB per square meter per day are not only much lower than the downtown grade-A office rents of 12 to 15 RMB, but also those of the officially developed clusters, which often ask between 8 and 12 RMB, well beyond the budget of design firms.[13]

In addition to cheap rents, communal spaces and amenities are provisioned with the designers in mind, rather than with a commercial outlook that many developer-produced clusters favor. The Warehouse Café is priced reasonably, so much so that it is cited by Timeout as one of the most affordable cafés. With fresh ingredients that highlight the environmental consciousness of the designer-owners, the café caters to both the tenants with similar affinities and like-minded visitors. It also accommodates gatherings and events that are often important parts of the design offices' repertoire, and is one of the many public spaces in the development designed for spontaneous meetings, encounters, and networking between different designers. On each floor meeting rooms are placed at prime locations, rather than stuck between desks, to facilitate these meetings. The recognition of the nature of public and shared spaces as crucial elements to design culture is being extended to the larger set of amenities that the Anken designers are planning for a site near the Warehouse. Fitness studios, for spin and yoga classes, would not only benefit the creative labor that comes but also reach into the community that is also changing in the area.

The recent completion of a skyfarm on the roof of the Warehouse building highlights the role new concepts play in the development of the spaces. Functions that are welcomed by the transnational set of creative workers— many of them from places like Brooklyn and the East End where the latest ideas about urban agriculture, low carbon–footprint organic vegetables, and slow food are being tested—are created to the sensibilities of the incoming creatives, rather than the other way around.

Just as Shanghai's beginnings as a modern metropolis in the nineteenth century were shaped by foreign direct investments (FDI)—both in the capital and human resources that became stakeholders for the development of the nascent city—so the re-opening of the city in the 1990s is also shaped by the influx of knowledge and capital that took on the local frameworks in a learning process that included international expertise. Especially in the city center the assemblage of capital, international expertise, and local access transformed the city from the drab, overcrowded, and run-down city of the 1990s to one filled with establishments that resemble those in any global metropolis. And despite the official high entry cost for small studios, many transnational start-ups are willing to set up an enterprise in Shanghai because of the openness to foreign ideas and the growing demand sophistication.[14]

More than half of the occupants at Anken are from an international background, either as expats or locals who have had stints abroad in training. The studios of Anken thus serve as nodes of knowhow exchange, filtering ideas from abroad and localizing them; and as a microcosm of the openness of Shanghai that has attracted many of the foreign talents to the city in the first place. The receptivity of the talent hub harks back to the cosmopolitan era of 1930s Shanghai, updating the city's legacy of openness in the contemporary shift of global trends.[15]

In contrast to the top-down curation by developers for creatives, the selection of the incoming tenants by the designer-developers is also seen by the designer-entrepreneurs as being of utmost importance. This fundamental aspect of community building and formation of an ambience rather than simple commercial co-location is also the reason that, different from many other creative clusters created by SOE-backed firms interested in short-term returns and producing oversupply of generic spaces, Anken Green has had good occupancy rates from its inception. When the space was nearing completion, proposals were submitted by both multinationals and local SOEs to become the ground-floor tenant of the most visible and accessible real estate of the building. The designer-developers rejected both the international conglomerate of DHL and the local SOE of Friendship Store. Both were financially secure and well-connected enterprises that would have ensured steady returns. But the wait and being selective about the right ground-floor tenant yielded a better fit. The Danish design store Paustian moved into the space in 2008. Paustian decided to base its representative for international markets in Shanghai to meet both the growing demand for exclusive designed furniture in the expanding Eastern market and the increasing utilization of production in Asia. The 300-square-meter showroom and sales office at Anken serves as an exhibition space for incoming clients as well as the site from which local customizations are designed. The furniture design showroom utilizes the high-ceilinged former industrial space, and showcases the types of products and services offered by the knowledge economy and transnational creative entrepreneurships in the building.

The music agency Massive Music was the first tenant to take over a space at the top of the six-story building. With offices in Amsterdam, London, New York, Los Angeles, and Shanghai, it is also representative of the kind of value chain that serves the growth and demand of the Asian market. Their clients include large multinationals like Nike, Audi, and Sony, who have also arrived in the expanding East Asian market. Meetings with representatives of creative sectors from around the world call for the kind of spatial ambience that Anken Warehouse supplies. It would take another year and a half for the building to be at full occupancy, but the selection of the right tenants is seen to be more important to the longer term commitment to the community than immediate returns.

Figure 3

Architects, graphic designers, photographers, and theatre consultants have settled in, some in the form of studios and others starting with renting by the desk while their business expanded. The diversity of spaces is designed with the growth of the successful start-up in mind. From individuals to a small, growing team to becoming commercially established creative firms, the design community offers camaraderie, through events and gatherings, that is beneficial to the tenants in terms of practical advice as well as more conceptual dialogues. Commercial registration for small offices is an added service for start-ups, many of whom would otherwise work in unregistered spaces. Although the official clusters promise a platform service, formal services are often charged at consultancy prices; the reality of informal networking far surpasses the fulfillment of "creative city" goals on paper.

Most small, private developers choose not to invest in such urban retrofit projects today because the returns are not worth the investment, unless the relationship to the state—both the district and the SOEs—is excellent and assured. More recently, the return on investment is usually around seven to eight years because of the shortened lease term of ten years from SOEs and higher rents since the initiation of the cluster as a business plan.[16] The uncertainty of being able to continue the lease also discourages a commitment to a space rather than seeing it as a contribution to the city.

Even though the Anken project on Huai-an Lu is only 6,700 square meters—officially not qualified to be honored with the "creative industry cluster" designation that requires at least 10,000 square meters—it was recognized because of its reputation for being able to attract desirable creative enterprises, large and small, and curate a certain kind of vibrant, conversant community. The officials at the district acknowledge the project's contribution to the city and consult Anken in the five-year plan to help shape the vision for the future of the district. Through the decade of working in the Jing'an district, the Anken designer-developers are familiar with the key decision-makers in the bureaucracy, and cite the district as one of the most progressive in Shanghai. Its personnel are under direct authority by the municipality, and the rotating leadership is made up of ambitious, young entrepreneurs, often well-connected with the outside and commercial world.

Since economic liberalization in the late 1990s, the leadership boldly rejuvenated the historically famed West Nanjing Road area through land marketization and attraction of overseas Chinese development capital to remake the district's thoroughfare into an alternate central business district (CBD). From the Portman Hotel complex to Plaza 66, the first international hotel, high-end commerce, and grade-A office spaces brought in the revenue that would help develop the other parts of the district. The fact that these much needed functions are centrally located and not on the greenfield developments in Pudong showed the audacious vision of a group of experimenters who gambled on a win. Given the locational advantages of Jing'an as central and accessible, a current plan for creating a specialized fashion sector involves not only retail but also schools and production spaces to make use of the labor force in design, marketing, and production. Involvement of prominent re-use experts like the architects Neri and Hu in the renovation of a former police station shows the forward thinking of the district council in integrating international concepts and localizing them in context to compete in the global framework.

The openness of the district leadership and ministry authorities is extremely important in the Chinese context of discretionary decision-making that is embedded in the relational planning. The differing physical and socio-economic conditions of the districts have marked out territorial competition among them that have also set up differing preferential policies in relation to creative development. By contrast, the outlook of the district of Huangpu was outdated; the imagined immediate and high returns from targetting high-end financial services industries were unrealized because of the unclear property rights of much of the former financial district's building stock, creating a barrier to efficient redevelompent plans. Anken's attempts to initiate change in the historic architecture behind the Bund have only met with frustration. The municipal level SOEs in Huangpu that are in perpetual possession of much property in Shanghai's city center[17] are holding out on their real estate assets, resulting in many properties that are otherwise prime locations in the city remaining empty.[18] And despite tax incentives by all the districts to attract creative cluster development, discrepancies remain in project realization due to the differing demand and thus prioritization of the district authorities.[19]

An alternate mode of cluster development by the designer-entrepreneurs of Anken Green, although officially recognized by the authorities, reveals the divergence of a cosmopolitan, user- and community-based approach to creative spatial production from the customary processes for creative cluster formation by developers. The importance of curation in the formation of a like-minded design network and the amenities that extend beyond the cluster enclave provokes a rethinking of the conventional business model for creative incubation. Taking the policy left over from the planned economy but innovating spatial reuse beyond the standard established by the official set of business plans, Anken and other small designer-entrepreneurs have offered alternative creative hubs that are not on the official map. Not unlike the initiatives taken by the artists of Moganshan Road, which has since become packaged to be the M50 cluster, the designers are forming their own community that goes far beyond that of the designation of a cluster into the transformation of the city.

Jing'an Villas: Bottom-up Clustering of Creatives

An upper-middle-class residential neighborhood block with a center axis that connects the busy commercial thoroughfare of Nanjing Road to the parallel Weihai Road, Jing'an Villas has undergone partial commercialization since the mid-2000s, like many city-center residences. What distinguishes the development in this area from the usual process of local residents renting out ground-floor spaces for commercial use under the tacit approval of the local government is the clustering of transnational creative functions in the *lilong* quartier. With a café called Chabrol that hosts weekly film screenings, small boutiques and designer showrooms, exhibition-cum-atelier spaces, a library, and services such as a spa that specializes in Israeli olive soap, the shop owners epitomize the local cosmopolitans that operate between multiple cultures.

Reinforced by the informality of the commercial conversions and the increasing difficulty to formalize conversions, signage advertising new enterprises appears only when the ateliers, boutiques, and cafés are open. With often irregular opening hours, the small businesses seem to disappear easily back into the residential fabric. Without knowing that a particular gallery or designer showroom is located down a specific side lane, one could easily miss it simply walking down the main lane, mistaking the neighborhood for any of the other increasingly few older residential quartiers that remain in the city's core. In the quiet bustle of everyday life, where older residents gather chatting down one lane and deliveries are being cycled to their destinations in another, the young, Airbook-toting freelance generation deliberately chooses Jing'an Villas for its kind of proximate seclusion for their fledgling enterprises. Only a stone's throw away is Shanghai's most famous commercial thoroughfare, Nanjing Road, where grade-A high-rise offices sit atop large-scale malls—built in the late 1990s by overseas-Chinese developers who first introduced Louis Vuitton to China—towering over the diminutive *lilong* house area. However, low rents in what continues to be the state-managed but partially marketized historic buildings in Jing'an Villas remain a practical attraction for the creative entrepreneurs.

Figure 4

Cheap rents in a central location are complemented by the quality of modern architecture and urban structure. Ground-floor units for commercial as well upper-floor residences in the heritage-designated quartier harbor an ambience that many of the creatives prefer as the habitat for their own design productions. The owner of Gezi Café, one of the earliest to move into a ground-floor unit with a front terrace entrance, emphasizes the quiet of the location and character of the new *lilong* compound. The graduated privacy of the public space network, in proximity to the bustle of commercial life at the periphery of the block while also preserving the quiet of the interior public spaces, shaped the social success of the urban morphology in both historical and contemporary reuse (Bracken, 2013). The atmosphere of the built environment from the 1930s is cited by the café owner by Gregory Bracken as what is suitable for the appreciation of his personally roasted coffee from different parts of the world. The revival of coffee culture in the context of economic liberalization, according to Hanchao Lu, "…is not just about coffee drinking per se, but about the city's presumably Westernized culture as many Chinese understand it." But here more than at Xintiandi or in Plaza 66, it is represented by the "children of reform (who) wish they had lived in their grandparents' age" (Lu Hanchao, 2002: 178).

The setting representative of the historic Shanghainese identity, one that is modern, westernized, and refined, also helps highlight a cultural emphasis on the goods and services of Jing'an Villas and the lack of a prevalent commercial flavor.[20] Distinguished from commodity items that are taking over creative quarters such as in the now fully commercialized Tianzifang, the products and services offered at Jing'an Villas are presented as unique, creative, and cultural. Tianzifang in the early 2000s is often referenced as being similar to the current state of Jing'an Villas in the early 2010s , with partial commercialization, low rents attracting young start-ups amidst residential functions, and creative entrepreneurs producing distinctive designs; however, Tianzifang has since become overwhelmed by commerce, with increased rents driving out the creative entrepreneurs.[21] In Jing'an Villas, the small spa that specializes in Israeli olive soap emphasizes the unique friendships and life-changing experiences that its owner brings to the services it offers visitors. Despite having expanded into other parts of the city since its humble beginnings, the importance of maintaining the original space tries to underscore the sharing of these experiences beyond that of commercial exchange.[22]

Anxiety about the potential for over-commercialization in Jing'an Villas, from the existing 30-70 ratio of commerce to residence to 70-30 in places like Tianzifang, defines the mindset of the young entrepreneurs for the changing conditions of doing business in Shanghai. Ever-ready to move and find new locales, the small creative entrepreneurs are prepared for the everyday effects of unpredictable and discretionary policy implementations on creative development in Chinese cities, whether state-backed promotion of a residential quartier's conversion into a creative commercial area, such as in Tianzifang, or short-notice functional change and closure, as exemplified by the shut-down of the neighboring Weihai Road 696, an art factory which housed many ateliers and galleries in an international mix.

Many young creative entrepreneurs have returned from stints abroad, some have worked for extensive periods and others are still working part-time in multinational creative firms before taking the plunge to open up their own enterprises. The decision to start their own business comes from both the desire to express individual design aspirations and from the interest to fill a market they often feel is being undersupplied by the existing choices.[23] A small tea shop owner opened her space as respite from the rat-race of the corporate world. A jewelry designer wanted a place to share his connoisseurship of pearls without the commercial pressure of having to push them onto customers.[24] Connections and experiences forged and garnered in multicultural settings are complemented by the necessity of being able to negotiate with the local resident committees and street offices for real estate procurement and commercial approval. Although often without the necessary commercial licensing that is obligatory for formal enterprises, small penalties are seen as a fair tradeoff for the low rent and flexibility that are afforded by being off-the-record, as long as there is no official disapproval. The combination of local agility with international perspective enhances the realization of global aspirations within the Chinese institutional frameworks.

Once inside the spaces such as Chabrol, one is reminded of international trend quarters such as Berlin's Prenzlauerberg or New York's Williamsburg. Visitors and consumers who manage to find the area are those who have access to the selected networks of Japanese and European design magazines or are in the particular type of expat circles whose patronage of the area emphasizes the "hidden" factor. As many design shops had online showrooms before they were able to offer physical showrooms, online publicity through channels such as Weibo, Facebook, and Twitter easily link the like-minded vanguards beyond the locality. The under-the-radar feel of the neighborhood's transformation, the young entrepreneurs' shirking of official designation as a "creative cluster," and their deliberate embedment deep inside a residential neighborhood contrasts it to the mode of development where a developer oversees the curation of the different commercial programs. The individuated outlooks are nevertheless compensated by a tight network of horizontally-linked small creative entrepreneurs. The self-formulated affiliations between the cafés, boutiques, spas, and galleries create an urban ecology that thrives on the perpetuation of progressive creativity and deliberate exclusivity. The 2010 shutdown of adjacent Weihai Road 696 that led to an influx of many of the young designers to the Jing'an Villas also united their collective weariness against state-appropriation and converging artistic aspirations.

One of the main reasons that the buildings of Jing'an Villas did not suffer the same fate as many other *lilong* did since the 1992 push to renew 365 hectares of existing old housing by 2000 through demolition and reconstruction is because of its architectural typology. In their design, new *lilong* types, as contrasted to the old *lilongs* such as those at Tianzifang and Xintiandi, are more adaptable to contemporary living and upgrades. Unlike the older types that require structural changes, including plumbing installation and parking accommodation, the new *lilong* types were already designed with modern plumbing and heating infrastructure (Zhao, 2004). The mere size of each residential unit makes reuse for contemporary functions more possible than in the older type of *lilong*, where upgrade and commercialization in entirety is necessary for the physical form of the units to be viably preserved. Especially in the western end of the French Concessions, the new style *lilong*, which has a wider arterial system with eight-meter-wide passages as compared to the narrow four meters in the old *lilong* type, were designed to accommodate the car. Car parking spaces and attached garages were also constructed as part of the design. The architecture is modern in scale, and the infrastructure, material, and style of the new *lilong* types thus enabled their contemporary reuse. The diverging architectural legacies of the old and new *lilong* types, often built in the same decade in the 1930s but contrasting in the qualities mentioned above, have also contrasted in their roles in the contemporary modernization of a re-globalized Shanghai.

The *lilong* house was originally designed for one family; however, from before 1949 population increases forced the division of the house into multiple units shared by many families. The ground floor was separated into front and back to allow a private entrance to the upper floors from the back, with the front entrance and terrace reserved for the ground-floor unit. With the exception of the intermediate floors, through which the residents of the upper floors must pass in order to reach their units, the house became separated into individual independent units, with private kitchens installed on the roof terraces. The demographic swell continued with no provision for housing construction since 1949, when housing tenure became severely fragmented and densities multiplied through the chaos of the Cultural Revolution. Amenities and infrastructures that were shared by multiple families became also the spaces of contestation.[25] It was during this time that the modern spatial provision for infrastructure became residential units to house the growing urban population. Under the planned economy, central heating was discontinued south of the Yangtze River, and boiler rooms were reused as dwelling space. Garages, similarly, became irrelevant and were reused. In these ground-floor spaces for infrastructure, programmatic changes are visible as the market economy returned.

It has often been reported that the creative transformations in the Jing'an Villas seem to trail the incremental bottom-up development at Tianzifang. Tianzifang and the Xintiandi project are often seen as setting important precedents in Shanghai and other Chinese cities in their successful reuse of historic structures for new commercial functions (Rutcosky, 2007; Su, 2008; Tsai, 2008). Tianzifang is individually developed by small entrepreneurs directly leasing from the district or the residents who still hold onto their property use rights (Zheng, 2011). However, Xintiandi underwent en-bloc acquisition by a single developer whose preservation of the old buildings for commercial leasing to high-rent tenancy is only a small percentage of the entire development, the remainder of which underwent and is still undergoing the demolition-reconstruction process (He and Wu, 2005; Ren, 2008).

Unlike Tianzifang's development, it is not only commercial conversion that is changing the quartier. Many of the small entrepreneurs also choose to live in Jing'an Villas for the same reasons that their businesses are started here. The modernity of the new *lilong* as typology makes it also feasible to convert old units into spaces suited to contemporary living without the need for expensive structural alterations. Loft-like spaces on the top floors and the layouts around the central stairway are characteristic of the typology and have made the buildings favored for their ambience, which is not offset by high prices. A Chinese-learning American musician who shares a roof-top flat with a self-taught landscape designer from Australia is among those active in the "scene" of the film screenings, pop-up events, and exhibitions, but without their own store on the ground floor. The convenience, the famous local noodle stall that is open daily at the end of an adjacent side lane, and the cheap rent of publicly-managed buildings also are key ingredients for the network of creative types who settle in Jing'an Villas neighborhood.[26]

In the late summer of 2013, the district government demanded the closure of all small enterprises. The conflicts caused by food and beverage joints and bars disturbing the neighboring and upstairs residents led to the eviction of all small enterprises on the grounds that they are illegal.[27] But as some residents reflected, the improvement of the built environment through privately funded renovations and landscaping by some of the small entrepreneurs cannot be overlooked.[28] And self-mediating mechanisms within the neighborhood had already led to the closing down of bars and restaurants that opened a few years prior, which have been replaced with quieter and cleaner enterprises such as art studios and knitting courses. Just as the Weihai Road 696 closure only two years prior left an abandoned property empty, the future of the ground-floor spaces is only certain in what is not allowed, rather than the potential of what they could be.

The coexistence of the local and the global, along with the typological resilience of the modern architecture, have made Jing'an Villas a decidedly convincing product of bottom-up reuse and upgrade. Localized cosmopolitans subtly harness the legacies of modern spatial structures and try to stave off full-scale commercialization, which would upset the balance struck between commerce and culture. The specificities of the legacy structures, both in architecture typology and urban morphology, are urban resources for the continued development of Shanghai as a global city.

Anfu-Wuyuan: Mixed-use Creative Neighborhood

Since the early-2000s, streets like Jinxian, Xinle, Fumin, and Julu Road in Shanghai's city center have turned into trendy areas with design shops, cafés, and restaurants. These streets with older architecture typologies and more rundown building stock are often the first to be rejuvenated on an individual scale by small entrepreneurs who not only individuate their creative reuse but also recover the rent gap that impending economic development brings.[29] Spatially, converted terraces of *lilong* houses, street-front ground-floor spaces of apartment buildings, and insertions into and constructions from garden walls in the early 1990s were used for everything from convenience stores to hair salons and small restaurants, originally necessitated by increasing demand that was no longer fettered by planned economy (Fig. 5). Some are still operated by local, small-scale entrepreneurs, initially forced into surviving the SOE reforms of the late 1990s through commercial enterprise (Davis, 1999). Others evolved into cafés, boutiques, design ateliers, and event spaces that are run by younger generations, a constellation of locals, returnees from overseas, as well as expats, hosting a range of events that link the international value chain to the locally situated spaces and producers.

Figure 5

Catalyzed by new high-rise apartments and the upgrading of the properties of the Drama Arts Center, Anfu Road quickly became the newest trendy street in the late 2000s. The success of the redevelopment clusters within walking distance, like Le Passage Fuxing and Ferguson Lane, both incrementally developed by transnational patrons who have settled in Shanghai since the early 1990s, also offered the commercial potential for catering to the growing expat community who choose to settle in the heritage-designated neighborhood.[30] Two properties on Anfu Road belong to the Drama Arts Center: the building that houses Mr. Willis, and Baker & Spice, spinoffs from the Wagas chain of fusion foods; and the converted former cinema that is now the Anfu Court complex with the imported food supermarket Pines catering to international clientele, Sunflour bakery, Y+ yoga, Dragonfly spa, and serviced co-working spaces. The Drama Arts Center is a state institution with allocated land whose decision for upgrade not only responded to the demands of the new occupants of the residential towers inserted at Anfu and Wulumuqi Roads,[31] but also propelled organic developments along the street, triggering a number of small international fashion, merchandise, and furniture designers to open shops in the ground-floor spaces of the *lilong* and terrace houses. Many began as informal commerce in low-rent spaces to test the reception of the changing market, but the growing demand sophistication of the rapidly shifting social composition ensured the short-term longevity of these programmatic additions.

Inside lane 249 of Anfu Road is a two-story garden house that hosts the atelier space of a young fashion designer; she shares the house with an eco-events non-governmental organization (NGO) and freelance real estate agents along with two residents, a yoga teacher and an English teacher. The house had been subdivided to house five families until the 1990s when one of her relatives took over a unit from a family who emigrated abroad. The designer took over one of the residential units, originally as a place to live, because of the convenience of the location and the low rent. Working in the public relations and event planning sectors, embedded in the transnational networks in Shanghai, she realized that like-minded people were looking for products that were often either undersupplied or overpriced due to high demand.[32] Like many other small entrepreneurs, she entered the niche market of product and fashion design as a consumer who detected a market demand. With an affordable base in the middle of a changing neighborhood that is populated with likely customers of similar taste, she turned her second-floor space into a working studio where fabrics are test cut and assembled before sending to a larger workshop to be produced in quantity. Despite the lack of renovation to the 1930s garden house, one in an ensemble of seven, the neighbors in the lane—including a retired local opera diva, a Taiwanese architect, a French gallerist, and the event planning group downstairs—offer the kind of networking connections that make the embedment in the area important for starting up her business.

Walking through the neighboring lanes that traverse the interior of large blocks bounded by the main arterials—characteristic of urban morphology of the western end of the former French Concession—ground-floor conversions into architecture, furniture design, graphic design, and advertising studios are visible from the renovated façades and signage. The initial startups are usually informal, drawn by the cheap rent and convenience of location. A part of the residential fabric, many are not yet registered commercially, not unlike those in the Jing'an Villas. But some have expanded from their ground-floor spaces into the upper levels and adjacent new structures. Others have shifted from hidden spaces catering to the creatives to become more commercial.

The bookshop 1984, for example, was initially tucked away on the ground floor of a garden *lilong* house with frontage on the street. The owner, who is not originally from Shanghai, is one of the many transplants who have since stayed; he worked in design and advertising while he opened his shop as a place where friends can gather. Because the commercial function is not official, the bookstore is not visible as such, aside from a small sign on the door that denotes the space inside. Only those who know can ring the doorbell to be let into the bookstore. Having a space in the Jing'an Villas, the design entrepreneur likes the quiet and "hidden" feel of the space and the ambience for the books and hosted events.[33] The small backyard that is part of the typology of garden *lilong* houses makes possible an extended open space that could host events, such as readings and movie screenings. But so as not to conflict with the upstairs residential neighbors, the space that has served coffee and small snacks closes at ten in the evenings.

Since the early 2010s, the terraces of ground-floor houses are being converted as street-front shops expand to Wuyuan Road, and the street is joining the latest vogue. Real estate agents who offer small boutique owners the trendy streets of Jinxian, Xinle, Fumin, and Julu Roads shifted their gaze onto the still under-recognized former market street. In the ground-floor of a house with a large garden on Wuyuan Road, the atelier and showroom of the Xinlelu platform for young local and international designers took over the space from their former partners Lolo Love, an events space and shop that specialized in imported European vintage fashion, accessories and furnishings. Unlike most commercial spaces facing the street, the showroom is entered via a lane leading from Wuyuan Road through the spacious garden, presenting the proximate, accessible, but sequestered ambience of the place. Although inward facing, the garden holds a number of movie screenings by local film-makers, exhibition openings by local and international artists, and community clothing drives. Among the co-owners is a Chinese-American designer from California whose store, William the Beekeeper, was the first vintage clothing shop in Shanghai, and a Shanghainese designer who did a stint in the UK and who lives and has a showroom in the Jing'an Villas.[34] As with many of the ground-floor commercial conversions, the house was originally one residence that has since been subdivided for many families. The fragmented ownership of the house itself is inviting small investors speculating on the future of the street to start buying up the "use rights"[35] from elderly neighbors living upstairs in the planned effort to eventually consolidate them. At the moment, it is precisely the fragmentation of ownership and the informality of the commercial space that has kept the rents low and made it affordable for designers. But as many expressed the uncertainty of the longer term future, the readiness to shift to the next location is ever present.

In the neighborhood around Anfu and Wuyuan Roads, the *lilong* houses that only a decade ago were home to multiple families have become studios where young designers, photographers, and architects share work, event, and exhibition spaces. The ground floor of a garden house is an atelier shop with fashion sourced from Berlin and Paris and a garden hosting photo shoots, film screenings, and salons. Garden vestibules have become charming sun-roofed cafés. Front parlors have been converted into shop windows. Along with minor commerce, small agile firms producing knowledge-based design services and products grew in the city center, transforming existing structures with innovative reuse and programming. It is not only in the historic and modern buildings but also in the post-war slabs and small manufacturing hubs that the architectural diversity of the neighborhood is nurturing the creative sectors. Underground bunkers have been turned into wine cellars and live music venues that host some of the best underground music in Shanghai, and *lilong* houses are becoming intimate settings for young galleries. The urban fabric exemplary of a Chinese modernity that is hybrid and cosmopolitan—low-rise, high-density, Bauhaus-planned and art-decoed *lilong* houses, Spanish turreted garden houses, and high-rises reminiscent of nineteenth century Paris or London, with workers' housing blocks, towers, hotels, and malls tucked in between along boulevards lined with plantain trees—is nurturing microcosms of globalized entrepreneurial experimentation and innovation.

Politics and Aesthetics of Creativity

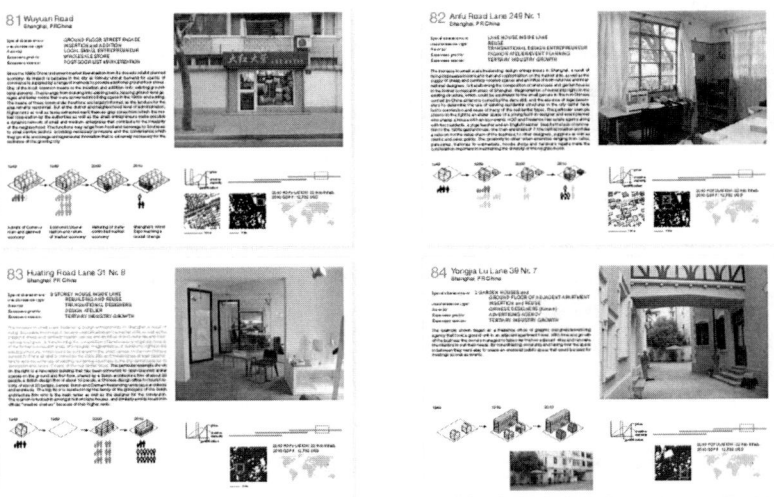

Figure 6

Returning expat Chinese, foreign city dwellers, as well as the new, local middle class demand and produce spaces of consumption and leisure as well as production and residence. At the same time, the older generation of locals staying behind find themselves hosting the grandchildren sent back into the city center to attend the more prestigious schools, filled now also with children of expats and returnees. The locals staying behind become landlords in the same way the villagers of the urban villages have within the last decade, as the local *hukou* hold the privilege to capitalize on the prime real estate locations. And the younger generation of locals with transnational connections and knowhow open up the cafés, trattorias, and boutiques that oblige the globalized palette of the latest "locals."

Conclusion

As in most post-socialist cities, the return of capital and demand generated spatial opportunity.[36] Unique to Shanghai as a former global city is the extremely rapid usurping of international trends in the grasping of these opportunities. Within a decade, bits of Shibuya, Prenzlauerberg, and

[80]

Williamsburg are filtered by the expat adventurer or foreign talent and by the Shanghainese who has returned from the stint abroad, to bring parts of Tokyo, Berlin, or New York. New influences are quickly adapted by the local market, whose cosmopolitan history and contemporary hybridity has made it the easy to promote imports. The constellation of returning diaspora compelled by both nostalgia and pragmatism, expats attracted by the city's historic global connections, and expediently learning and commercial-minded local stakeholders are not only facilitating the re-entry of Shanghai into the global market but are becoming the active residents and users of the city center area, which has been shirked by the local residents reminded of the pre-reform decades of forced socialization and privacy deprivation.

Even as the local families of the city center look forward to new lives in the towers outside the first ring road, spacious homes with new-found privacy and infrastructure are sold as part of the commodity housing package, and newcomers settle into the romanticism of the former Concessions area, untainted by shadows cast by the Socialist era on the center cityscape, expats and Generation Y'ers even nostalgically admiring the references of the Cultural Revolution era. "Pop cultures from Hong Kong, Taiwan, and America are becoming decisive influences in the formation of a new urban culture. A new pop culture has arisen from the combination of these imported influences and the almost 'exotic' rediscovery of traditional culture after decades of interruption" (Hou, 1996: 167).

The predominant rhetoric for clustering comes from the discourse of competitiveness promoted by business school professor Michael Porter that features agglomerative effects as crucial to the knowledge transfers and synergies that produce the added value of new economies (1998). The scale of clustering in space is embedded in a creative ecology that is intimately tied to the urban habitat in which creatives work and live. In the three vignettes presented, from the scale of the building, to the block, to the neighborhood, the functional mix that allows for adjacencies and proximities to enhance living beyond working is crucial to the everyday development of the creative city. The small-scale enterprises, much more agile and flexible, are faster than the bureaucratic apparatus and decision-making processes, and are incrementally upgrading the city center. With tacit understanding of urban attributes beyond those of "new" and "big" in global cities, they have expedited the process of Shanghai regaining the status of a world-class city. Their proficiency in cultivating the spatial qualities of legacy urban structures is important if not crucial as a marker for a maturing knowledge economy. And more critically, their engagement of the local institutional frameworks may offer alternatives to the adaptive governance that prevails in carrying out of urban rules.

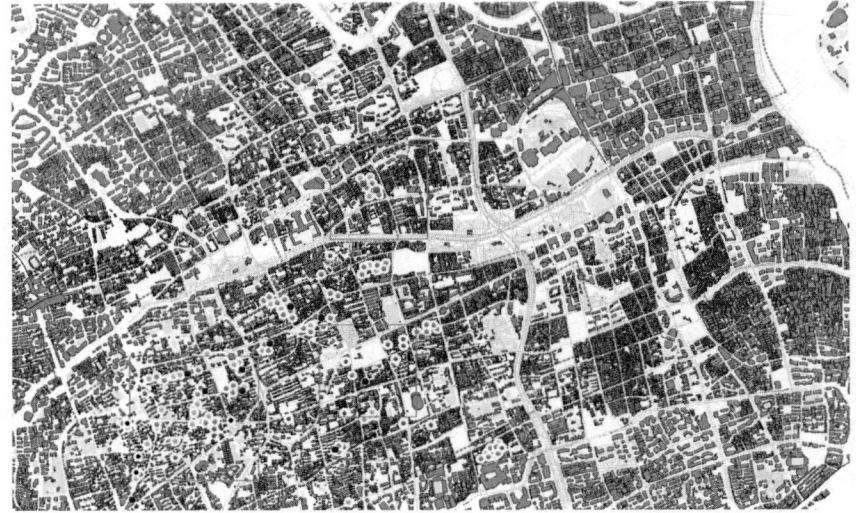

Figure 7

Notes

1. Reuse of former industrial quarters by artists as studios and for exhibitions is a well-known phenomenon in Western cities. The form it takes on in China, of course, is not only related to de-industrialization but a fundamental restructuring of the socialist units of production.
2. The leadership of the Creative Industries Center even admits that only 10 percent of creative productions come from the official clusters.
3. There is a noted disconnect between the study of the modern historic structure for the sake of cultural heritage and the reality of its contemporary market demand. At the same time, upgrade efforts also bypass the diverse residential and commercial demands, which are often in conflict, to inject programmatic pastiches of Shanghai's former cosmopolitan representation; Lu Hanchao offers an optimistic cultural dissection of this (2002).
4. *Lilong* housing is the typical high-density, low-rise lane house typology produced in the beginning of the 1900s in Shanghai.
5. Twitter is officially banned in China, so the "tweeting" is done via an extremely popular parallel site called Weibo. For the lack of better word, the Western version is used until weibo'ing becomes an official English verb.
6. The New Urbanists react to the automobile-driven proliferation of sprawl by the endorsement of pedestrian-friendly mainstreet developments in American cities, where community-building and neighborhood ambience are spatially engaged.
7. Interview with the Anken developers, 2013.
8. Policy from the late 1990s allowed the development of allocated land by the SOEs, with the profits realized to resettle workers. But the policy tightened in the early 2000s when formal processes for transfer of status from allocated to lease-hold were required to control the

over-commercialization of former allocated land through low-cost transfers (Tang, 2013).
9. The "three non-changes and five changes" *sanbubianwubian* policy issued by the Shanghai Creative Industries Center in 2005 states that there is to be no change to building structure, meaning no development by demolition-reconstruction; there is no change to ownership, meaning the SOE retains its perpetual leasehold on the state-allocated land; and there is no change to land-use status, meaning officially land use remains industrial where "creative industries" as a unique ambiguity between service and manufacturing is condoned and encouraged. As Jane Zheng reports in her research of Creative Industries Clusters, one third of jobs in the clusters are not related to creative industries (2011: 13).
10. The year 2008 also saw the release of a central government directive on encouraging the development of service industries, which was followed by a municipal directive on development of the modern service industry in Shanghai (Tang, 2013).
11. Until the mid-2000s the building was owned by the municipal level SOE Shanghai Great Wall Biochemical Pharmaceutical Factory, and used as a storage facility. The building, built in 1933, is one of the first mechanized slaughterhouses in the world, and its modern concrete form marks it out as a unique piece of heritage architecture for Shanghai. But despite its architectural importance, the curation of tenants for its reuse is visibly unsuccessful in its vacancy. This, coupled with its legacy architectural qualities, has led to a business plan of high-rent spaces that makes the building more event-oriented than used on an everyday basis.
12. Interview with the Anken developers, 2013.
13. The official creative cluster of 1933, for example, charges 12 RMB/sm/d, in 2013.
14. In Shanghai, many small design firms are run by expats who came to China in the mid-2000s, even though competition with the large design institutes is increasingly dominating the design landscape and the cost of starting a business is expensive: one million RMB of registered capital to start business in China. In Singapore, for example,

starting a business is initially less expensive and procedurally much easier; however, few small creative studios are transnational because of higher expenditures, limited creative ecology of other creatives, and the narrow breadth of the geographical market.

15. Shanghai's beginnings as a concession city had, in the early twentieth century, cultivated a bevy of locals from the region whose openness and adoption of foreign, largely Western tastes, mannerisms, and lifestyles was first the envy and later the scourge of the city. Whether they had hailed from Ningbo or Yangzhou, their mastery of the "modern" (according to a 1920 letter to the French Concession municipal council of Shanghai from the archive, "modern" is synonymous with "European") and their ability to adapt the "modern" in the local Chinese context would become the outlook and skill for the many who came to live in the early twentieth century "special economic zone."

16. Interview with the Anken developers, 2013.

17. The dominant SOE conglomerates are Shang Tex (textiles), Paper Factory, Shanghai Electric, Bailian (including N. 1 Department store, Yong'an, Hualian), and Shanghai Automobile Group, which are often so large and fragmented that one branch holding on to property in one location has no knowledge of other properties in another location, frustrating innovations in reuse and rejuvenation by small creative entrepreneurs.

18. The SOE leadership often set up local companies to manage the properties, waiting to be compensated at market price should the district government want to requisition the state-allocated land. This holdout is not much different from the expectations of the local residents holding on to the use-rights of their residences should re-development come with compensation, even if they no longer live in their original abodes. Budget hotel chains, for example, are often developed on these state-allocated lands that cannot undergo redevelopment via demolition and reconstruction.

19. Interview with the Anken developers, 2013. Usually 40 percent of the tax goes to the district and the other 60 percent goes to the central government. For the startup period of two to three years, the return of

the 40 percent by the district as a tax break to the leasee was offered to attract developers of these creative clusters.

In an interview with another local developer who worked on several re-use projects in the Xuhui district, some of which have been designated "Creative Industry Clusters," the comparison of Xuhui to Huangpu was made to prove the point of necessary curation. Huangpu District offers incentives to attract the right kind of tenants, but Xuhui, being a smarter district, ends up with the higher class ones. Whether the comment indicates the demand sophistication of the district leadership or inherent structural advantages of each of the areas it matters less than that of the resultant impact on contemporary development.

20. Interviews with several small entrepreneurs, 2011 and 2012, show concurrence in their choice of the historic lane houses as venues for their creative commerce and the importance of Shanghai's refined cultural asset to their cultural credibility.
21. The small entrepreneurs who were interviewed in 2011 and 2012 shudder at venues such as Tianzifang and consider it to have become too commonplace and commercial.
22. Interviews, 2012.
23. Many of the young entrepreneurs originate from the advertising industry and diversify into creative productions. As Lu Hanchao noted in his article, Shanghai's revival of commercial advertising, regarded under socialism as an instrument of capitalism's gratuitous allure to consumption, shows the cosmopolitan legacy effects on the contemporary return to commercialism. Within this context, advertising as a service industry also simultaneously identifies market niches that are undersupplied, creating possibilities for entrepreneurial innovation.
24. Interviews, 2012.
25. The impact of the legacy conditions, in particular from the post-1949 period, on contemporary developments in Shanghai is often largely overlooked (Zhou, 2013).
26. Interview with some of the creative entrepreneurs and residents also yielded a network that connects to other hotspots along the creative conversions.

27. From official reportage of the procedure for evicting the small entrepreneurs, the district has taken the chance to consolidate decentralized bureaus and build a platform on the example of Jing'an Villas. See http://www.jingan.gov.cn/newscenter/zxft/bndft/201310/t20131022_134070.html.
28. Interviews, 2013.
29. The concept of a rent gap as the differential between realized value and potential value was one of the causes identified by Neil Smith for the process of gentrification (1979). In the post-socialist context of Chinese cities, where real estate values and their locational differentials were obliterated under the planned economy, it could be said that the rent gap propels the development of the city.
30. The growing emphasis of built heritage in reaction to the 365 Plan that designated the goal of demolishing 365 hectares of dilapidated housing for reconstruction by 2000 ushered in the establishment of the Historic Building Conservation Law in 2003, followed by the 2004 designation of twelve "Historic Cultural Styles Districts" *lishiwenhuafengmaoqu* .
31. In a plan drawn up in 1939 by the French Concession government that outlined building regulation, the area on which the residential towers Kingsville (1997), Chevalier Place (2004), and The Summit (2006) are located was a pre-French Concession urban settlement, an "urban village" at the corner of Anfu and Wulumuqi Roads, around which modern typologies would later be permitted to be built, according to the strict zoning regulation. Because of the area's pre-modern typologies, infrastructural provisions, and narrow network, it seemed destined for destruction. In the face of development pressures in the 1990s, selective replacement for urban upgrade led to the oldest part of the concession area being earmarked for demolition; because of age, it did not and could meet modern standards of city building.
32. Interviews with the designer-entrepreneurs, 2012.
33. Interviews with the owner, 2012.
34. Interviews with the designer-entrepreneurs, 2011 and 2012, revealed that small investors are paying elderly residents a lump sum for their "use rights" in neighboring buildings with the prospect of converting to "ownership rights" in two to three years. The elderly residents, often

with children abroad who therefore will not contest the ownership relinquishment, benefit from the immediate cash flow.

35. "Use right" is a uniquely Chinese definition of ownership tenure that marketizes the residential units, especially in historic city core areas, and allows the buying, selling and exchange of residential units between those registered with the local *hukou*; but it is different from the "ownership right" that comes with newly built commodity housing or apartments and whole houses or lane houses that have not been subdivided from the original typology. Once the "use rights" of all residential units have been slowly bought up, the possibility of converting subdivided "use right" into consolidated "ownership right" is possible. The limited supply of "ownership right" units, coupled with the growing recognition of the embedded cultural value of historic architecture since the mid-2000s, means the high demand for those units in the city center make their possible consolidation and conversion very lucrative.

36. This is visible in the cities of the former Eastern bloc in Europe as well as in the cities that marketized following Arab socialism, making the study as relevant in the larger context of understanding urban production systems.

Bibliography

Bracken, Gregory. *The Shanghai Alleyway House: A Vanishing Urban Vernacular.* New York: Routledge, 2013.

Christiaanse, Kees. "Urban Breeding Grounds." *Credit Suisse Global Investor* 2, no. 10 (2010): 25.

Davis, Deborah S. "Self-Employment in Shanghai: A Research Note." *The China Quarterly* 157 (1999): 22-43.

Florida, Richard. *The Rise of the Creative Class: And How It's Transforming Work, Leisure and Everyday Life.* New York: Basic Books, 2002.

He, Shenjing, and Fulong Wu. "Property-led redevelopment in post-reform China: a case study of Xintiandi redevelopment project in Shanghai. *Journal of Urban Affairs* 27, no. 1 (2005): 1-23.

Hou, Hanru. "Towards an Un-Unofficial Art: De-ideologicalisation of China's Contemporary Art in the 1990s." *Third Text* 34 (1996): 24-39.

Hsing, You-tien. "Land and territorial politics in urban China." *The China Quarterly* 187 (2006): 575-91.

Keane, Michael. "Great adaptations: China's creative clusters and the new social contract." *Continuum Journal Media Cultural Studies* 23, no. 2 (2009): 221-30.

Kong, Lily. "Making sustainable creative/cultural space in Shanghai and Singapore." *The Geographic Review* 91, no. 1 (2009): 1-22.

Li, Ming. 上海历史文化风貌区保护和更新的品牌策略 "Branding strategies for protection and renewal of Historic and Cultural Areas in Shanghai—a case study on Hengshan Road-Fuxing Road Historic and Cultural Area." 上海城市规划 *Shanghai Urban Planning Review* 3 (2012): 22-6.

Lu, Hanchao. "Nostalgia for the future: the resurgence of an alienated culture." *Pacific Affairs* 75, no. 2 (2002): 169-86.

Porter, Michael. *On Competition*. Boston: Harvard Business School Press, 1998.

Ren, Xuefei. "Forward to the Past: Historical Preservation in Globalizing Shanghai." *City and Community* 7, no. 1 (2008): 23-43.

Rutcosky, Kori. "Adaptive reuse as sustainable architecture in contemporary Shanghai." Master's thesis, Lund University, 2007.

Smith, Neil. "Toward a theory of gentrification: a back to the city movement by capital, not people." *Journal of the American Planning Association* 45 (1979): 538-48.

Su, Nanxi. "Art Factories in Shanghai: Urban Regeneration Experience of Post-industrial Districts." Master's thesis, National University of Singapore, 2008.

Tang, Zilai. "The renewal of allocated industrial land in the perspective of property right system: the case of Hongkou District, Shanghai." Paper presented at the conference "Institutions of property rights and sustainable Asian urbanization," National University of Singapore, 2013.

Tsai. Wai-Lin. "The Redevelopment and Preservation of Historic Lilong Housing in Shanghai." Master's thesis, University of Pennsylvania, 2008.

Wan, Yong. '上海旧区改造的历史演进，主要探索和发展导向 "Historical Evolution, Main Study and Orientation of Planning on the Redevelopment of Urban Old Areas in Shanghai." (Published in Chinese.)城市发展研究 *Urban Studies* 11 (2009): 97-101.

Wang, Jun. "Art in capital: Shaping distinctiveness in a culture-led urban regeneration project in Red Town, Shanghai." *Cities* 26 (2009): 318-30.

Yang, You-ren, and Chih-hui Chang. "An Urban Regeneration Regime in China: A Case Study of Urban Redevelopment in Shanghai's Taipingqiao Area." *Urban Studies* 44 (2007): 1809-26.

Yeh, Anthony Gar-on, and Fulong Wu. "The New Land Development Process and Urban Development in Chinese Cities." *International Journal of Urban and Regional Research* 20, no. 2 (1996): 330-53.

Zheng, Jane. "'Creative Industry Clusters' and the 'Entrepreneurial City' of Shanghai." *Urban Studies* 48, no. 16 (2011): 3561-82.

— "The 'entrepreneurial state' in 'creative industry cluster' development in Shanghai." *Journal of Urban Affairs* 32, no. 2 (2010): 143-70.

Zhou, Ying. '上海中心城区：在全球愿景和本土构架之间 "Between global aspirations and local frameworks: city center Shanghai." 城市中國 *Urban China* 56 (2013): 68–73.

Zhong, Sheng. "From Fabrics to Fine Arts: Urban Restructuring and the Formation of an Art District in Shanghai." *Critical Planning* 16 (2009): 118-37.

— "Production, Creative Firms and Urban Space in Shanghai." *Culture Unbound: Journal of Current Cultural Research* 4 (2012): 169–91.

Zhu, Jieming. "Local developmental state and order in China's urban development during transition." *International Journal of Urban and Regional Research* 28, no. 2 (2004): 424-47.

— "Urban Development under Ambiguous Property Rights." *International Journal of Urban and Regional Research* 26, no. 1 (2002): 41-57.

CHAPTER THREE:

In Praise of the "Coffin": Urban Sociality in the Japanese Capsule Hotels
Non Arkaraprasertkul

Introduction: Roland Barthes in Tokyo

If I want to imagine a fictive nation, I can give it an invented name, treat it declaratively as a novelistic object, create a new Garabagne, so as to compromise no real country by my fantasy (though it is then that fantasy itself I compromise by the signs of literature). I can also—though in no way claiming to represent or to analyze reality itself (these being the major gestures of Western discourse)—isolate somewhere in the world (faraway) a certain number of features (a term employed by linguistics), and out of these features deliberately form a system. It is this system which I shall call: Japan (Barthes, 1983: 2).

To many, Japanese cities are full of surprise. In his famous essays later compiled and published as *Empire of Signs* (1983), the renowned French semioticist Roland Barthes wrote about how much Japan fascinated him when he first visited in the late 1960s. Barthes raised questions such as "Why do Japanese people write from right to left—and top to bottom?" "Why do they bow; and who bows to whom?" Among the most famous questions he posed was "Why is the center of Tokyo empty?" These questions were indeed exciting to Western audiences, especially in the late 1960s when Japan had just hosted the Olympics, projecting again its influence to the world at large. These questions, if they were to be asked of the Japanese both then and now, would probably be met with equally puzzling answers: "How else should we read if not from right to left—and top to bottom?" "Why shouldn't we bow?" And, of course, "Why shouldn't the center of Tokyo be empty—that's where the Imperial Palace is." We can only take Barthes' questions only moderately seriously as he neither claimed to be an expert in Japanese culture, nor wanted to say that his view was absolute. In 1970, when *Empire of Signs* was originally published, it was the champion of cultural studies during an age of post-structuralism. His thesis about Japanese society was meant to reinforce the fashionable poststructuralist notion of semiotic instability; that is, there is no such thing as concrete meaning. His aim was to create a dialogue about the ways in which we could understand the systems of signs differently. His achievement in doing so was unquestionable: his comparison was based on the Western notion of things, which was why *Empire of Signs* was well received, because it spoke to the basic "common sense" of the majority who read it.

The Capsule Hotels

Had Barthes not died an untimely death about a decade after *Empire of Signs* was published, no doubt he would have returned to Tokyo many times. Although he did not admit that he loved Tokyo, anyone reading the book would know that he was deeply fascinated by the city. If he were to go back today, he would certainly not be able to avoid or overlook the significant role of the *kapuseru hoteru*, literally the "capsule hotels," in the city today. On average about seven feet long, four feet wide, and three feet tall, these coffin-like boxes stacked on top of one another are in fact spaces where people must crawl in to sleep. There is nothing more than just that space with some basic amenities such as a small, built-in television and an electronic alarm clock, and a bonus: some air inside to breath. Once you are in the capsule, you cannot do anything but sleep. The height of the capsule is just enough that you cannot sit up straight. The width of it is just small enough that you would not be able to rotate your body full-circle. You will be cut off from the world, at least visually, because all three sides around you will be walls that are less than two feet away your face. The capsules have been designed to maximize utility, vis-à-vis saving space, and, according to my own experience, the most comfortable sleeping position in one of them is to lie on one's back.

Although seemingly cut off from the world, this does not mean that clients will actually be in aural peace. In these hotels, each capsule has only a thin bamboo shutter in the front (through which you would crawl into your capsule), and thin plastic panels on both sides that separate your body from the corridor and your neighbors' capsules. If you are really unfortunate, you will hear snores all night long (or at least until you fall asleep). This is why newer capsule hotels provide pre-installed noise-cancelling earphones. If Barthes would visit a capsule hotel today, he may ask questions like, "Why would anyone sleep in a coffin?" "Why would anyone pay for the lack of privacy?" "Why would anyone do this to themselves?" My guess is that, for a celebrated French theorist like him, the quest to find answers to these questions would certainly not include his staying in a capsule hotel. Moreover, I imagine that he would be disgusted with it, and would come to a conclusion similar to his conclusions for all the things that shook his understanding of a semiotic system when he previously visited Japan. As many critics have said, he probably would "invent his fictive Japan" to explain why the Japanese did what he thought they did.

Who stays in capsule hotels? There is a serious dearth of academic research and publishing about the capsule hotels. The available literature touches upon the capsule hotels slightly, usually only to illuminate points about other topics (e.g., Chaplin, 2007; Leslie, 2006; MacDonald, 2000; McNeill, 2008). According to popular perception, the majority of the clients are those who miss the last train, which is around midnight, to go home. Given that a taxi ride back to many satellite towns where most office workers could only afford to rent cost around US$100, many who miss the trains, of course, choose to sleep in the city instead of paying a large taxi fare to drag themselves home just to come back again in a few hours in the morning. The second group of clients is the un- or temporarily-employed; for them, capsule hotels are often the cheapest housing option available (as they can rent by the month for about 1,000 yen, or about US$10, per night). It is also important to note that there seems to be an interesting gender dimension inside these capsule hotels. As I will show in my observations and interviews, most people I met were male, at least the customers. That said, some receptionists of the newer capsules I visited were women in their early 20s. According to the managers of the capsule whom I interviewed, it seems to be a trend to make the capsule hotel "more welcoming" by having women at the reception. When being asked why there were no women in capsule hotels, the popular responses from the customers were "because women have to go home to take care of the house, the child (or the children)." What this type of statement implies is that "men like us are those who have to work hard to pay the bills, including working overtime and staying in a capsule hotel." Although this highly masculinistic and solipsistic view expressed by the male customers might hold some truth, I will show in this paper that there are also other reasons why customers choose to stay in the hotels. Capsule hotels are not *de jure* exclusive to men, yet, *de facto*, they only serve men as there are no separate facilities for women (and by law women and men cannot share space in a public bathtub). In this paper I will intentionally disregard this specific point about gender and sociality due to insufficient research on the capsule hotels for women, and only focus on the typical capsule hotels for men.

Before I began my study of capsule hotels, all I heard about them from the media and those who have rented one is that "it's horrible." It is not surprising that all of these sources were friends, students, colleagues, and relatives who, in their entire lives, had never experienced anything smaller than a standard twin-bed hotel room prior to their encounters with the capsule hotels. The popular explanation for why the Japanese invented the capsule hotels is commonly rooted in the nation's unequal economic development, which I will later try to demystify.

Through ethnographic methods, I argue that the capsule hotel is not just a place of necessity. In fact, it is a very dynamic social space, as the main spaces in capsule hotels are not the coffins in which one sleeps, but the bathhouse, lounge, TV room, massage chairs, and so on. It is these elements alongside the playing with the boundaries that make it a dynamic social space—clients seek comfort from this space where the boundary between the private and public sphere is most unclear. My fieldwork consists of multiple and intermittent stays in these hotels in Japanese cities since my first stay at a capsule hotel in 2006. I will focus on three capsule hotels with which I am most familiar in two busy districts of Tokyo: Shimbashi and Kabukichō. Each has its own unique selling point, and the price for each also varies (the one in Shimbashi being the most expensive and the one in Kabukichō the cheapest).

A Brief History

The creator of the first capsule hotel was one of Japan's most renowned architects of the twentieth century, Kisho Kurokawa (1934-2007), who, throughout his life, was known for his ability to "peep into the future" to derive his creative energy (Koolhaas and Obrist, 2011; Ouroussoff, 2009). His designs and approach to architectural problem-solving were often future-oriented; 1979 was the peak of a new movement to reinvent architecture, based on the biological function of the building to grow, known as the "metabolism movement." Kurokawa proposed a series of "capsule projects" or buildings that consisted of temporary housing units that would, in his own words, serve "the salarymen or international businessmen who work late and would need only a small area to rest and rejuvenate before heading to their next destination" (as cited in Koolhaas and Obrist, 2011). The most famous capsule building, the Nakagin Capsule Tower in the bustling Ginza district in Tokyo, consists of decent sized rooms (7.5 feet x 12 feet x 6.9 feet), and is still standing today (Lin, 2011). Hints of the coffin-sized capsule hotels that he later developed can be seen through his earlier designs of furniture: all in white, all built-in, and all extremely economical.

Figure 1: Kurokawa's Nakagin Capsule Tower. Photo: Author

Kurokawa's "Capsule Inn" opened in Osaka in 1979 and is still in operation—the prototype of coffin-sized capsule hotels today. Some believe that Kurokawa got the idea for the coffin-sized capsule boxes from cardboard and plastic box shelters that the homeless and day laborers in the rundown Nishihari district of Osaka used as housing. These cardboard huts were pretty much the same size as his original capsule hotels, as both were designed for fundamental necessities, and hence just large enough for the size of the body. However, since the Nishihari cardboard huts did not really come into use until after the economic crisis of the late 1980s, the arguments that Kurokawa's designs of his first coffin-sized capsule hotel in 1979 were inspired from Nishihari does not really hold water. Consequently, I believe that it was Kurokawa's vision for the future lifestyle that brought to life the capsule hotels that we know today.

That said, it could well be the case that the capsule hotel entrepreneurs were inspired by the cardboard huts in Nishihari as a way of supplying cheap housing after the economic downturn of the late 1980s, which was precisely the time when the market for the capsule hotels began (Tanaka and Yamada, 2007). In Nishihari there were multiple "flophouses" for the day laborers to rent cheaply. Those capsule hotel entrepreneurs saw the opportunity to extend the existing market by offering cheap accommodation for the large number of commuters. The success of the first capsule hotel gave rise to the next, and now they are almost ubiquitous in all business areas in big cities in Japan (Boonbanjerdsri, 2012; Leslie, 2006). Since the late 1980s, Tokyo and Osaka have been two of the most expensive cities in the world. Like any other cities whose primary mode of production is tertiary, real estate is at the center of all forms of investment (Sassen, 2001). Unless they already owned a house or an apartment before the economy took off in the late 1980s, middle-class Tokyo residents would find it almost impossible to afford to buy or rent a room in the center of the city near their workplaces (Cybriwsky, 1991), hence the capsule hotels have since then become a popular choice among urban workers.

According to the geographer Roman Cybriwsky (1991), the suburbanization process of the 1980s combined with, what he calls, "the squeeze on inner neighborhood(s)" gave rise to many urban innovations such as the capsule hotels, love hotels (also see Chaplin, 2007; Jacob, 2008), and urban coffee shops (also see White, 2012), among others. With the advent of a modern train system, the local government of Tokyo decided to push residential neighborhoods to the outer rings of the city. While there were many benefits to such ideas—for example, larger open spaces for people, much more affordable "4LDK" houses (a house with four bedrooms plus living room, dining room, and kitchen) and "2LDK apartment" for the lower-middle-class, and so on— the suburbanization process has since created a clear spatialization of class (Freedman, 2011). Although the famous modernist architect Yoshinobu Ashihara (1986) is, by and large, quite optimistic about this concept of "bed towns," he was also skeptical about the concept when it comes to the matter of equal access to resources and facilities. Such resources and facilities seemed to be made only for those who could afford to live in the city's center. Because the main business area is still in the center of the city, the white-collar, middle-class urban workers have to commute from the suburbs and outskirts, as well as from nearby prefectures to the city center to work. Sometimes, the length of commute could be as much as two to three hours, one way. Those who have traveled by train from the suburbs to the city center will be quite familiar with the scene of the morning train with everybody sleeping. The journalist Amy Chavez (2013) even calls it "a rite of passage into Japanese society."

Many *sararīman* (literally "salarymen"), the ubiquitous suit-and-tie white-collar workers whose lifestyle revolves entirely around work at the office, usually work around the clock (Vogel, 1963). Many of these workers live in the suburbs or in nearby vicinities of Tokyo as they cannot afford to rent anywhere within the city; this explains the appeal of these hotels, which allow the workers to eliminate an otherwise long commute after a late night of working. Further, the capsule hotels, as I was first told by a colleague who was researching Japanese history at the time, is a "retrofitting mechanism" that was invented by an opportunistic entrepreneur precisely to solve this problem (Arkaraprasertkul, 2010). Imagine this: if you live three hours away from the city center, going to work and returning home takes six hours, which is one-fourth of your day. Why should you go home just to have two to three hours of sleep and then get back on the train again? The idea behind the capsule hotels is: Why not just give people what they need—enough place to sleep, because that's all they need? In contemporary Japan, the structure of urban employment in Japanese major cities leaves little room for those who are not salarymen since big corporations are running the major part of the economy (The World Bank, 2013). In fact, many capsule hotel clients arrive at the hotel late at night, put their belongings in a personal locker, take a shower in a shared bathroom, with an option to hang out afterwards in a room full of massage chairs, before climbing into their personal capsules and going to bed.

Hence, from the historical perspective, capsule hotels are nothing more than, simply put, coffins for temporary sleeping. The hotels only provide what an unconscious body needs, and unconscious bodies do not mind if they are stacked on top of each other. The questions here are: first, just because we are so used to larger rooms does not mean that spaces in the capsule hotels are too small, does it? In other words, how do we understand the capsule hotels in their own terms? Second, are the capsule hotels simply affordable private spaces in the city? Apart from the capsules themselves, the majority of spaces in the building that house a capsule hotel are public areas, such as the bathhouse, locker room, massage room, and so on. These public spaces serve about one hundred to two hundred clients every night. How then, is it possible that the capsule hotels are just simply space for individual isolation? Do they have any social functions?

Ethnographic Encounter: A "VIP Suite" in Shimbashi

In the following three stories a window is opened onto a space that is rarely discussed by academics relative to capsule hotels. It is this space that I experienced through my ethnographic study. At about ten o'clock at night, I checked into a capsule hotel in Shimbashi, a major interchange station on the east side of Tokyo and a ten-minute walk from the famous retail district of Ginza. It was my Japanese colleague who recommended this hotel because it was new and very close to the Shimbashi train station. "So, you can't miss it," said this friend, and like he said, I encountered the hotel after leaving the station and walking only about five minutes through a small lane full of cafés, bars, and noodle stools. The capsule hotel "Peace" (pseudonym) was located in a seven-story building that, from the exterior appearance, must have been a typical 1960s' concrete tower before it was turned into a capsule hotel.

Figure 2: The author in a typical "coffin" capsule. Photo: Ken Takahashi

Automatic glass doors slid open, and an instantaneous blast of cold air from the air-conditioners hit my face. Upon entering, I was welcomed by walls of shoe lockers that stood directly in front of me and a sign that the no-shoe zone started here. I put my shoes in one of the small lockers, closed it, and took the key to the counter where two male receptionists greeted me. I was fortunate to have with me Nobu, a recent graduate from a college in Tokyo, whom I knew through a good friend of mine; Nobu assisted me with translation. In his early twenties, Nobu himself had never stayed in a capsule hotel. He grew up in the suburbs of Tokyo before coming to the city for college, where he stayed in a college dorm for the last four years (in fact, in the same room all the way through!). His comment (which I will return to) was rather interesting: "I am not a salaryman; why would I ever stay at a capsule hotel?" Translating for me, Nobu told me to give the shoe-locker key to the receptionist in exchange for an actual room key, which was attached to a watchband-style bracelet for safe-keeping around my wrist (because I would need it to get in and out of the capsule compound to go to the bathroom and so on). The shoe-locker key would be kept for me until checkout. I gave one of the receptionists my credit card in exchange for a receipt (4,800 yen per night), and a brochure showing what was available to me: the amenities of the "VIP suite," for which I paid a premium.

I was pointed to the locker in which I would put all my belongings. In the locker, there was a towel, a green and white shirt, and three-quarter pajamas with the logo of the hotel on both pieces of the garment. I changed to that uniform pajamas uniform, put all my clothes and belongings in the locker, closed and locked it, and walked to the elevator with an electronic key to my room around my wrist. Though very curious, Nobu decided not to stay for a capsule hotel experience that night because there was still enough time for him to get on the train home. Yet, before he rushed to catch the last train, he translated for me the brochure that showed the amenities in the hotel:

> "An elevator is the only way to go up and down the building. The sentō (public bath) is in the B2 basement—that's where you'd go first for a bath (shampoo, rinse, and body wash were provided). Then, you'd go to B1 for a lounge area, where there were massage chairs, and "cubes" in which you could use Internet, watch TV, or listen to music that you like. There you'll find a vending machine from which you can get all kinds of free drinks. Once you're done, you'll go up to the fifth floor—the VIP suite—and find

In Praise of the "Coffin"

your capsule. Remember, make sure the number of the capsule is the same as the number on your key. I'll see you in the morning!"

Surprisingly, although the majority of spaces in the capsule hotel are public, clients treat it as a home both practically and ritualistically. As Ashihara writes in *The Hidden Order*, the sequence of activities is exactly the same: beginning with one taking off one's footwear before entering the house, leaving the dust and dirt at the gate. Then, one would go take a ritualistic bath and move on to the bedroom, the most private and least seen part of the house.

Although people did not converse in the public bath or sentō, which was always relatively quiet, they did when they came out of the public bath area to the communal basin where they brushed their teeth, cleaned their ears, and shaved. In many capsule hotels, the locker and the sentō are on the same floor, so that you do not have to put on the clean pajamas just to take them off again when you enter the sentō. It could be that this particular hotel was originally an office building that was recently renovated and modified to serve as a capsule hotel, so the infrastructure was not designed properly to be one. Once I came out of the sentō, I took an elevator just one floor up to the lounge area, in which, to my surprise, I found people in the hotel's green and white hotel pajamas watching TV, surfing the Internet, reading Japanese comic books of manga, playing video games, playing chess, and sitting in massage chairs.

The lounge was not big. The center of the room was a table, around which there were a few very comfortable chairs. Above the table was a big, flat-screen TV where, at almost all times, a baseball match was on. Next to the table were three gigantic vending machines: one for unlimited free soda, hot and iced coffee, and tea; one for instant noodles; and one for beer (the last two were not free but coin and banknote-operated). Around the periphery of the lounge were about a dozen "cubes," personal cubicles around three by five feet in length and width and about four feet tall, in which one would find a personal massage chair equipped with a fifteen-inch personal TV, a computer screen where one can get on the Internet, and headphones. In one corner of the lounge there was a small space for a personal massage, separately partitioned by floor-to-ceiling curtains. This personal massage area was open until 3:00 a.m., and there was always someone in there having a massage from a professional masseuse.

The lounge area was always full of people. I stayed up until 6:00 a.m. with strangers who flowed through the same space in the lounge area continually from when I first arrived around midnight. A couple of people left around 2:00 a.m., leaving me with about a dozen people until the sun came up. It was precisely this lounge area that got me thinking about the sociality of capsule hotels. If a capsule hotel is just a place for businessmen working or partying late who missed the last train home and needed a place to crash, what were these people doing in the lounge? Shouldn't they be sleeping in their personal capsules?

Hiro: A Salaryman

Hiro was a married salaryman in his early forties. His hair was half grey, but it was just enough to make him look graceful and respectable. Because he walked tall and dressed neatly, I thought he was a CEO of a company until the minute I saw him checking into the Capsule Hotel Dream. He told me that he had a house in the suburbs about two hours away from Shimbashi. Every day he had to change the train three times to get to the station closest to his home, where he would pick up his bicycle and ride about twenty minutes to reach home. After working eight to ten hours, plus about five hours of commuting time every day, he was always extremely exhausted when he got home, and was still very tired when the alarm clock went off, usually only about four hours after he closed his eyes at night. It was not that late at night when I talked to Hiro—about 11:00 p.m.—so I asked him why he did not go home since the train was still running. Because I was expecting a simple answer, such as "It doesn't make sense to go home just to get up four hours later," I was surprised when he shared his somewhat personal rationale, "I don't want to go home. Sometimes my wife doesn't really understand me," said Hiro.

It seemed to me that he was not comfortable with his socioeconomic status, and was frustrated with all kinds of pressure from his wife and daughter. "I have a teenage daughter who will demand more from my very little income, and my wife who doesn't like what I do very much," said Hiro. "Whenever I go home, we have nothing else to talk about except how she 'thinks' I should try to earn more money for the family." This was a striking comment about the current state of Japanese familial affairs.

From our conversation, it seemed to me that Hiro knew that he would not be able to fulfill the needs of his family—almost as if he was ashamed of himself—and for that reason, "a very long-hour workday, train schedule, and capsule hotel" were three perfect excuses for him not to go back home and face his wife and daughter. According to the anthropologist Merry White (2002), in a typical Japanese nuclear family, the role of the father has changed from the "thunder father," who exerts a strong sense of authority and makes all decisions, to that of an individual who sometimes only has limited authority (perhaps primarily economic?) due to the effects of occupational structure. The story that Hiro told exemplifies such change, as he believed that his wife and daughter did not really respect him because of his low income. He stayed in the capsule hotel about two or three times a week. Although it was quite cheap (about 3,000 yen per night), it was still cheaper to go home with a monthly train pass. "A night here is cheap but it does add up," he said, "but I'd like to live here forever if I have the money…why not?" Hiro always stayed at the Capsule Hotel Dream because of its "Stay ten nights get a free night" loyalty program. When staying at a capsule hotel, he usually spent half an hour in the hot tub and a full hour "hanging out" in the bathhouse, devoting about forty-five minutes cleaning every part of his body. He always stayed up late watching TV in the lounge. "It's good to be alone…I couldn't watch anything at home because our house is too small and my wife and daughter are usually annoyed by the noise," said Hiro. He enjoyed watching TV and drinking free, hot tea. I did not see him socializing with anyone in the lounge; he was basically just enjoying himself. He usually did not go to bed until 3:00 or 4:00 a.m. When I asked him whether he felt comfortable sleeping in such a small space, Hiro replied, "I have no problem with it at all, I actually like it because it is small. It's good to be alone. I *feel at home* here." The need to be alone was central to Hiro's affection for capsule hotel life. It seems counterintuitive to think that someone would feel at home in a capsule hotel, especially when that person did have a real home to which he could go. He got up around 8:00 a.m., which gave him enough time to brush his teeth, get dressed, and walk about eight minutes to his office.

Ken: "In the Cocoon I feel I was myself again"

I met Ken, a struggling musician in his late twenties, at the Capsule Hotel Orange in Kabukichō, Tokyo's red light district close to the world's busiest train station of Shinjuku. The Capsule Hotel Orange had been around for twenty years. It opened in the early 1990s, and had been making a profit ever since. The location was the main factor why it had been so successful: it was close to the business area of Shinjuku where many salarymen work, and it was in Kabukichō where they all came to get drunk after work. Most of the Capsule Hotel Orange clients were the drunk salarymen who partied so late that they missed the last train. Beyond the shoe locker area where I took off my shoes and put them in the locker, there was a front desk. "You need to buy a ticket first?" said the receptionist who then sent me back to the "ticket machine," a small machine with multiple buttons saying "short stay: one hour," "short stay: three hours," "overnight: checkout 8am," "shower only," and a few more. I put money in the machine and tapped "overnight: checkout 8am." A small piece of paper with the details of the reservation came out of the machine. I walked back to the front desk and gave the ticket and the key to my shoe locker to the receptionist, who in exchange gave me a towel and a set of sleepwear.

I usually call him Ken-Kun, instead of just Ken. (Kun is an honorific that the Japanese use to address a younger man denoting some familiarity between the two speakers. By calling him Ken-Kun, I felt that we were not complete strangers.) Ken-Kun thought of himself as a rebellious youth. You could see this from his dyed "Japanese punk" hairstyle and his clothing (jeans jacket and ripped jeans), which looked like he just came out of Woodstock in 1969. He obviously did not want to follow the societal norms, and did not continue on to college after he graduated from high school. Consequently, his parents, who lived in the suburbs of Tokyo, had been worried about him ever since. "I didn't want to bother them, and I love playing guitar, so I left home to do my own thing," said Ken-Kun. Unlike in other East Asian countries such as Korea, Taiwan, and China, university education is available to everyone in Japan because there are many colleges of different ranks. But because Japanese college students take college as "moratorium space" between the entrance examination and the company life, graduates view college education as the area of the least study in Japan's education system (Goodman and Phillips,

2003). No matter how hard they worked in college, graduates will have to be trained again by the company for which they will be working upon graduation. Hence, higher education is usually viewed as the "four years to have fun." Ken-Kun did not see the point in going to college. "I could have fun every day being out here. Look at (Japan's most famous novelist today Haruki) Murakami, he went to college and hated it." Ken-Kun resented the system.

Figure 3: An interior view of a typical capsule floor showing capsules stacking on top of each other in a typical low-ceiling type floor in Tokyo.
Photo: Author

A capsule hotel's "room" is usually long but narrow to save space. Ken's slim guitar case fit perfectly into the narrow locker, but then he was not able to hang up anything else; hence, he piled other things on top of his guitar case and reorganized upon checking out. Through a network of fellow young musicians, he would get phone calls from a pub, bar, or event, to go play about three or four nights a week; and given that his earnings were minimal, he could only stay in a capsule hotel. Ken-Kun was also talented with the piano, keyboard, and drums, thanks to his parents who could afford a private tutor to teach him music when he was growing up in the late 1980s. He had a true passion for music and thought that he could have done well in a music school. Of course, Japan's economy went downhill from the early 1990s onwards, so his parents could no longer afford his private lessons and, therefore, began to encourage him to think about the future in a big corporation where his social stability was guaranteed. This could be the reason why he rebelled against the system: it was almost that he felt "betrayed" by the system and his parents for not letting him pursuing his dream.

The lounge area of the Capsule Hotel Orange was more vibrant than that of the Capsule Hotel Dream because the neighborhood was full of pubs and karaoke bars. The hotel was located on the seventh, eighth, and ninth floors of a commercial building, so many clients, after they were done with their meals and usual after-work party with their colleagues, came up to the hotel to continue drinking with their close friends. The lounge was right next to the public bath, so many came out from the bath and walked directly to the lounge for a bottle of sake. Ken-Kun and I chatted there from around midnight until 5:00 a.m. There were more than twenty people in the lounge when we came in around midnight. By the time we left, there were about half a dozen people still chatting, another half a dozen of them fast asleep at the massage chairs, and about a dozen who had just come up for a cup of morning coffee. When we saw the sun come up behind the rows of tall buildings visible from the hotel's windows, both of us decided it was about time to "call it a night."

In Praise of the "Coffin"

Figure 4: A typical view of an entrance to a capsule hotel showing shoe lockers and self-service machine where customers pick the service they want before heading over to the reception desk for keys. Photo: Author

Ken-Kun did not always stay at the Capsule Hotel Orange, as he moved around the city to play music, but mostly the venues were around Shinjuku and the nearby areas. Like Hiro, he cared about the loyalty program that the hotel offered ("Stay eight nights get one free night," suggesting more competition in this area). But unlike Hiro, the only reason Ken-Kun stayed at the capsule hotel was because it was the only place he could afford. It was true that he could have gone home and stayed at his parents' home, yet the fact that he did not fulfill his parents' expectations and the role of a good child by going to college in order to "get a good job in a corporation, bank, or major conglomerate," just like both of his parents, made him socially vulnerable. He did not tell me much about his family life, but he said he would try to see his parents once a month. They usually wanted him to stay overnight whenever he visited, and he wanted to stay with them, but usually he would get a call to go play somewhere; and for a struggling musician in his financial situation, he could not turn down an offer even to play children's songs at a kid's birthday party. When I asked him how much he enjoyed staying in the capsule hotel, he replied, "It is nothing like home, and I would never call it a home. One day, I want a home where I have shelves for all of my records and a good stereo system. But there is one thing about the capsule hotel: the small space just for me. Whenever I crawl inside the "cocoon" (his personal nickname for the capsule), I feel protected. I have always been driven by other people; they push me around. It's noisy out there. But in the cocoon, I feel like I was myself again."

When Japan's most revered modern novelist Jun'ichirō Tanizaki (1977) talks about the beauty in the absence of the excess, he refers to the methods of meditation, which enable one to see one's real need. I think it was reasonable to see the absence of spaciousness here as meditation. Because once you are in the capsule, and the bamboo shutter has been pulled down, all you see is blank white space no matter which way you roll your body. There were just blank walls (this would be where Barthes' quote "...nothing more, nothing else, nothing" is most appropriate! [1983: 50-1]). You are lying in a space that provides nothing that you do not need: it was a space for meditation. Cut off from the world and from the expectations of society, the young Ken-Kun could feel like he was himself: this was what Tanizaki views as the beauty of the absence. Yoshinobu Ashihara also made a similar comment: the Japanese sense of aesthetics is all about the appreciation of the interior, as opposed to the exterior (1986: 100). The underlying structure of the capsule room, from this perspective, is very similar to a typical Japanese bedroom space, which is usually an empty tatami (traditional mat) floor, walled-in from all sides by plain paper walls.

Unlike Western-styled bedrooms in which there are many objects, such as side tables, a powder table, a television, bookshelves, and sometimes even working desk, Japanese bedrooms are just a space for sleep. There are storage spaces for linen and blankets built in to the walls of the bedrooms and concealed by sliding panels. However, when it comes to bedtime, all sliding panels would close, turning the walls of all sides of the room into simply solid, clean, and blank vertical plains, enclosing the space for sleep—nothing more, nothing less. It is the space for sleep as meditation.

Another aspect that was rather intriguing about Ken-Kun's comment is that he felt he was "protected" when he was inside the capsule. One of the major concerns that non-Japanese tourists have when they stay in the capsule hotel is security. There is no door; all capsules are separated from the corridor by only a vertical bamboo shutter. One of my colleagues posed the question, "How do you know that there won't be any crazy person invading your capsule when you sleep, robbing or even killing you?" This was a legitimate question: how do you know the person sleeping next to you is not a mentally unstable person? You wouldn't go to bed without having locked your door, hence why would you sleep among complete strangers without having any door at all?

The capsule hotel's security system is both active and a system built on trust. By law, every floor has a closed-circuit TV, and inside the capsule there is an emergency button. Yet, as Ken-Kun told me, "What the clients really need to know is that in the capsule hotel, people watch out for each other." Ken-Kun shared with me that although he never encountered anything untoward in the capsule hotel, he would always keep his eyes open for any unusual activities. In my own experience, apart from the lounge and public bath areas, I barely saw anyone. Once in the capsule compound, I knew that there were people staying in most of the capsules because most of the capsule's bamboo shutters were down, but I did not see them. I had no idea who they were or how long they had been in the capsule. There were noises (mostly snores), but otherwise the capsule compound was a rather quiet place. I had to walk very softly because walking loudly was considered inconsiderate, let alone making noise or talking on the phone (I had to go to the elevator hall to do so).

Frameworks for Analysis

To prevent myself from oversimplifying Japan —as Barthes did— I resort to two cultural frameworks to help me build my argument about the unique forms of urban sociality found in a place where one would not necessarily expect it. The first framework is Tanizaki's *In Praise of Shadows* (1977 [1933]). Tanizaki is one of Japan's master novelists whose profound sensitivity toward the, broadly defined, "Western and Japanese cultures" enables him to offer critical comments about the different ways of thinking about aesthetics among the Japanese and others. Aesthetics, for Tanizaki, is not simply a matter of beauty, but efficiency and functionality. In fact, one can conceive of *In Praise of Shadows* as an antithesis *to Empire of Signs*: unlike Barthes, who made instantaneous remarks about what he saw as unique in Japan, Tanizaki meditated on those differences. In *In Praise of Shadows,* Tanizaki shows us that there is also a socio-temporal element in the semiotics of aesthetics: "We (Easterners) will immerse ourselves in the darkness and there discover its own particular beauty. But the progressive Westerner is determined always to better his lot. From candle to oil lamp, oil lamp to gaslight, gaslight to electric light—his quest for a brighter light never ceases, he spares no pains to eradicate even the minutest shadow."

In other words, it is the shadow that allows us to see the "depth" of things, and it is always in the depth of things where the true beauty of life lies. While Western semiotics focus on the system of signs only as they are to be viewed in clear light, the Japanese sense of aesthetics is determined by the interplay of light and shadow, absence and presence, darkness and lightness, and so on. For instance, when Barthes looked at the interior design of a traditional Japanese house, he made the comment, "Turn the image upside down: nothing more, nothing else, nothing" (1983: 50-1). The panoramic photo of the interior view on which he comments is a typical corridor; on one side there is a paper wall that separates the interior from the exterior, and on the other side there is a series of sliding doors leading to the inner living room and the alcove of the house. The photo was taken in broad daylight in which you could barely see the depth of field. It was also composed symmetrically, hence the corridor in the middle looks particularly flat, overwhelmingly repetitive, and simplistic (i.e., per Barthes' comment: "nothing more, nothing else, nothing"). It was Barthes' choice to use such an image to discuss his view of a Japanese interior, which would receive a completely similar take from Tanizaki:

> Whenever I see the alcove of a tastefully built Japanese room, I marvel at our comprehension of the secrets of shadows, our sensitive use of shadow and light…The "mysterious Orient" of which Westerners speak probably refers to the uncanny silence of these dark places. And even we as children would feel an inexpressible chill as we peered into the depths of an alcove to which the sunlight had never penetrated. Where lies the key to this mystery? Ultimately it is the magic of shadows.

We can see here that the two logics are different. What I think best summarizes the essence of Japanese aesthetics in Tanizaki's eye is the architect Louis Kahn's comment, "The sun never knew how wonderful it was until it fell on the wall of a building" (Tanizaki, 1977). What *In Praise of Shadows* demonstrates, both literally and metaphorically, is that in evaluating the Japaneseness of the capsule hotel, we must take into account the absence as much as the presence of its physical elements. It might lack spaciousness, but in such absence there is a profound gain. After only a few nights, I could feel the beauty of this small space.

Secondly, the architectural theorist Yoshinobu Ashihara sets out *The Hidden Order* (1986). Ashihara's analysis reveals the deeper structure—hidden order—of modern Japanese aesthetics in all scales from an object to a house, from a house to a garden, and from a garden to a city, which essentially lies in its simplicity despite the complex appearance. Moreover, it is the flexibility of Japanese socio-urban infrastructure that enables its cities, such as Tokyo, to expand, morph, and change over time according to urbanization's demand. Hence, as much as one would love to think that the capsule hotels are an ad hoc mechanism for a city that urbanized too rapidly in the past three decades, it might well be the case that the capsule hotels were fundamentally Japanese. That is, Ashihara is astute to point out, perhaps if we strip down all forms of the contemporary urban—the capsule hotels, for example—we would see the same basic structure of Japanese dwelling culture, which never changes.

The capsule hotels are rather cheap, which has to do with the high demand rather than marketing strategy, as there are also other types of hotels. For instance, the business hotels (a small room with a private bathroom) are equally cheap, but they do not give you the social experience. The capsule hotels provide clients with meditative as well as social spaces. I show in my ethnography that sleeping is not always the only goal of the clients of the capsule hotels. Just like modern cafés, clients see the lounge space of capsule hotels as the place where they can see and be seen, but not be required to have any interaction with anyone. There is a need to get away from personal difficulties, and the physical facilities of the hotels, such as the public bath and lounge, provide the clients with both ritual and mental emancipation. The economic and social pressure makes their actual home feel unlike home; hence, the capsule hotels serve as a temporary home for those who need rejuvenation. By being surrounded by strangers whose sense of being, belonging, and personhood is defined by similar social factors, one could feel "at home." Clients hang out on massage chairs and drink in the lounge space, where they can leave behind the reality of their difficult situation, before crawling into the small coffin-sized capsule that simulates a private universe for the clients. The space inside each capsule is just small enough for one to feel that one need not worry about any other responsibilities. The beauty of the space is in the absence of the excess. These are the unique qualities that make the capsule hotels an important social institution. Just like the way Tanizaki and Ashihara praise the shadows and the hidden order, *I praise the coffin*.

Conclusion

In this chapter, I have tried to understand the capsule hotels in the context of contemporary Japanese society. I began by referencing Roland Barthes, whose *etic* view (general, non-structural, and objective in its perspective) of Japan had for decades shaped the impression of the non-Japanese. I aim to contrast his view with an *emic* perspective—or the understanding of a particular phenomenon in terms of its internal elements and functioning—by using frameworks provided by two well-known Japanese cultural theorists, combined with research on modern Japanese society. Through my own experience and the personal stories of several individuals staying in these hotels I show how these spaces can challenge; I show how unique the capsule hotel is as an institution. If we look beyond its appearance (form) into its actual use (function), the capsule hotel exemplifies the essence of Japanese architecture that not only serves the changing purpose of urban life, but also adapts to the new forms of urban social change. Tokyoites treat the capsule hotels as their urban oasis: a space of hiatus in the city where everything is moving at a breakneck pace.

Acknowledgements

This chapter has benefitted greatly from a series of conversations with my undergraduate students in the Spring 2013 "Societies of the World 33: Tokyo" class at Harvard University, Cambridge, Massachusetts. I would like to thank Stéphane Emmanuel Fouché, Jennifer Allison Klein, Taylor Morris, Dae Lim, Sarah Elizabeth Orlando, Christopher Cole Walleck, Blake Thomas Walsh, Blake Andrew Sundel, and Paul T. Yarabe. My deepest gratitude goes to Professor Theodore C. Bestor, my academic advisor and mentor, for giving me the complete freedom and opportunity to explore this particular research topic, and use the findings in teaching the class. I also owe my readers and colleagues Leo Pang, Lu Pan, Xinyan "Sunny" Peng, Trude Renwick, and Tobias Zuser for reading the entire manuscript and making extremely constructive comments throughout. Last but not least, I would not have been able to conduct this research at all without the kind help and warm support, especially in terms of translation, from my friends from Tokyo: Ken Takahashi and Tomomi Goto. *Dōmo arigatō gozaimashita!*

Bibliography

Arkaraprasertkul, Non. "Crypto-urbanism: retrofitting Tokyo." *Journal of Urbanism: International Research on Placemaking and Urban Sustainability* (2010): 127-29.

Ashihara, Yoshinobu. *Hidden Order: Tokyo through the Twentieth Century.* Tokyo and New York: Kodansha, 1986.

Barthes, Roland. *Empire of Signs.* New York: Hill and Wang, 1983.

Boonbanjerdsri, Kimberlee. "Capsule homes: creating space within space." Bachelor's thesis, Massachusetts Institute of Technology, 2012.

Chaplin, Sarah. *Japanese Love Hotels: A Cultural History.* London: Routledge, 2007.

Chavez, Amy 2013 Sleeping on the train — a rite of passage into Japanese society. http://www.japantimes.co.jp/community/2013/03/09/our-lives/sleeping-on-the-train-a-rite-of-passage-into-japanese-society/#.VJYSHUAtw.

Cybriwsky, Roman A. *Tokyo: The Changing Profile of an Urban Giant.* London: Belhaven, 1991.

Freedman, Alisa. *Tokyo in Transit: Japanese Culture on the Rails and Road.* Stanford: Stanford University Press, 2011.

Goodman, Roger, and David Phillips. *Can the Japanese Change their Education System?* Oxford Studies in Comparative Education, Vol. 12 (1). Oxford: Symposium Books, 2003.

Jacob, Ed. *Love Hotels: An Inside Look at Japan's Sexual Playgrounds.* Lulu.com, 2008.

"Kisho Kurokawa: Nakagin Capsule Tower Building." *Designboom.* 18 November 2011. http://www.designboom.com/architecture/kisho-kurokawa-nakagin-capsule-tower-building/

Koolhaas, Rem, and Hans-Ulrich Obrist. *Project Japan: Metabolism Talks.* Cologne: Taschen, 2011.

Leslie, Thomas. "Just What Is It That Makes Capsule Homes So Different, So Appealing? Domesticity and the Technological Sublime, 1945 to 1975." *Space and Culture* 9, no. 2 (2006): 180-94.

Lin, Zhongjie. "Nakagin Capsule Tower: Revisiting the Future of the Recent Past." *Journal of Architectural Education* 65, no. 1 (2011): 13-32.

MacDonald, Rob. "Urban Hotel: Evolution of a Hybrid Typology." *Built Environment* 26, no. 2 (2000): 142-51.

McNeill, Donald. "The hotel and the city." *Progress in Human Geography* 32, no. 3 (2008): 383-98.

Ouroussoff, Nicolai. "Future Vision Banished to the Past." *New York Times.* 6 July 2009. http://www.nytimes.com/2009/07/07/arts/design/07capsule.html?pagewanted=all.

Sassen, Saskia. *The Global City: New York, London, Tokyo.* Princeton: Princeton University Press, 2001.

Tanaka, Yukio, and Tetsuo Yamada, dirs. "Mapping the Future, Nishinari." Zakka Films, 2007.

Tanizaki, Jun'ichirō. *In Praise of Shadows.* [1933 in Japanese] English translation. New Haven, CT: Leete's Island Books, 1977.

Vogel, Ezra F. *Japan's New Middle Class: The Salary Man and His Family in a Tokyo Suburb.* Berkeley: University of California Press, 1963.

White, Merry I. *Coffee Life in Japan.* Vol. 36. Berkeley: University of California Press, 2012.

— *Perfectly Japanese: Making Families in an Era of Upheaval.* Twentieth Century Japan: The Emergence of a World Power Vol. 14. Berkeley: University of California Press, 2002.

The World Bank. "Doing Business in Japan." The World Bank Group, 2013.

CHAPTER FOUR:

Rediscovering the Japanese Houses in Taiwan: A Contest between Postcolonial Inhabitants and the Creative City Regime
Shu-Mei Huang

Taste of Colonial History

In December 2011, an exhibition titled "Japanese House" by the London-based Japanese photographer Tomoko Yoneda was launched at the ShugoArts gallery in Tokyo. The series of photographs was previously shown at the 2010 Kuandu Biennale to express "Memories and Beyond." Through the gaze of the artist, the Japanese-style residences built in Taipei under the Japanese occupation (1895-1945) became artistic objects in the artist's postcolonial representations of places linked with her country's recent imperial history. The kind of gaze is not the privilege of the artist, though. Over the past decade, Japanese houses have become significant subjects in historic preservation in Taipei. Today, they accommodate some of the most popular cafés and restaurants. Not only urbanites but also international tourists frequent them, especially those from Hong Kong and Japan. Most people, nevertheless, would not spend as much time reflecting on the unraveling histories in Japanese houses. To the consumers who randomly take in the hybrid narratives offered by the growing cultural tourism in the city, it is probably the taste—the taste of the cappuccino as well as the wooden sliding door frames—that differentiates those cafes from Starbucks and Hard Rock.

The trend has spread beyond Taipei. "The Japanese house"[1] has become a popular theme in public culture and creative industries in Taiwan. Among the rich heritage of Japanese colonization, the particular clustering of Japanese houses appears to be a construct layered by emerging interest in adaptive reuse of historical buildings and creating cultural quarters as the object of urban governance.

Figure 1: The Japanese House in Taipei. Taken by Shu-Mei Huang in 2012

Most of those small houses, nevertheless, were part of a larger establishment of institutions and industries associated with imperialism. But they are usually objectified and fetishized as particular subjects with beautiful black roof tiles and decorative elements as unique motifs. The complexity of colonial modernity is mostly twisted, covered, and transformed by a selected representation of taste, aesthetics, and memories to fit into a making of Taipei-style cosmopolitanness for the present. This chapter aims to examine the rediscovery of the Japanese houses and the capitalization on their presence in the city. To the artist Yoneda, "Japanese Houses" are not simply historical artifacts. They represent spaces in which modern and historical events are mixed.[2] The mixture, however, encourages a fetishism of colonial aesthetics coupled with a tendency to reduce creative industry into culinary business. The phenomenon can be observed in numerous Japanese house-turned-cafés and "creativity quarters" occupied by restaurants and bars.

Behind the fad was the city government's effort since 2004 of listing some of the relatively intact or important Japanese houses as monuments or historic buildings.[3] There have been thirty-four cases listed by the Department of Cultural Affairs at Taipei City Government to the end of 2011, especially from 2004 to 2007 (Table 1). One of the officials of the department proudly pointed out that "the wooden structures make the preservation of historical Japanese buildings a challenge, and even in Japan, most old buildings were destroyed or dismantled to make way for urban renewal projects" (Mo, 2009). In areas where Japanese houses are densely concentrated, such as Qingtian Street, Cidong Street, and streets near National Taiwan University, the colonial imprint on the urban landscape is now considered as "a moment of authenticity in some versions of commodification and at times in the scopic regime of the tourist gaze" (Keith, 2005: 128).

Table 1. The listing of Japanese houses as monuments or historic buildings in Taipei City

Year of listing	Number of cases
2004	10
2005	0
2006	8
2007	14
2008	1
2009	0
2010	0
2011	1
total (as of Dec. 2011)	34

*Adapted from source provided by the Cultural Bureau of the Ministry of Culture

There also has been a civic effort that continuously promotes the values of old, vernacular houses. In the city of Tainan, the non-profit Foundation for Historic City Conservation and Regeneration has successfully drawn attention to the cultural capital of old houses through a series of competitions from 2008 that awarded best practice for creating "New Lives in Old Houses." More and more old houses were successfully transformed into trendy bars, restaurants, and hostels, which quickly led to speculation over old houses. In Taipei, the Department of Cultural Affairs partnered with the Institute of Historical Resources Management in launching the "Old houses movement" to encourage private involvement in managing historic properties, many of which are Japanese houses.

The Wistaria Teahouse is probably the earliest example of adaptive reuse of a Japanese house before the building was listed as a monument in 1997. As the Australian writer Linda Jaivin recalls, "For some years, even as other teahouses sprang up in imitation of Wistaria around the city, Wistaria remained the meeting place of choice for many of Taiwan's cultural figures. I remember going there with the singer-songwriter Hou Dejian and our mutual friend the essayist Shu Guozhi, among others" (2012). Starting before the 1987 lifting of martial law, writers, dissidents and artists had met at the house even before it was officially run as a teahouse by Chou Yu. The teahouse represented an "alternative space" in a democratizing society. Rather than its architectural style, the unique socio-cultural history qualified the teahouse to be designated as a monument. The teahouse was saved so it could remain as a unique "teahouse salon" in the city rather than as an elegant Japanese house of colonial aesthetics (Jaivin, 2012). Nobody could foresee the fad for Japanese houses in the following decade.

This chapter is based on long-term research of the history of preserving Japanese houses in Taiwan over the past decade. The mixed-method research adopts participant observation, archival research, and discourse analysis to study the social construct of Japanese houses and the changing politics of historic preservation in Taiwan. During the years 2001 to 2003, I participated in the study of the Wistaria Teahouse with colleagues from National Taiwan University (Liu, 2003a). From 2004 to 2013, I observed the emerging interest in preserving Japanese houses. Lately, my participation in preserving the Colonial Taipei Prison Settlement ensures that the analysis is attuned to the changing politics of colonial heritage.

This chapter is divided into two parts. What follows the introduction is an overview of the history of preserving Japanese houses in Taiwan. It sheds light on the objectification of Japanese houses within a broader context of urbanization and industrialization. Meanwhile, it lays the groundwork for discussing the politics of heritage in which, as Gregory Ashworth reminds us, there is always a chance that we are colonizing the future with our current values (2009). The second part, through the case studies of "Huashan 1914 Creative Park" and the Colonial Taipei Prison Settlement, investigates the inconvenient relationship between colonial heritage and the construction of creative quarters. Engaging postcolonial thinking, I examine the creative city regime and its attempt to take over colonial vestiges in the city. Finally, I conclude the chapter by calling for more creative ways to restore postcolonial sensibilities.

The Recent Rediscovery of Japanese Houses in Taipei

Heritage has become a common paradigmatic example of gentrification and cultural economy in recent scholarship of postmodern cities (Zukin, 1987: 129-47; Jacobs, 1996: 16). With heritage preservation, cities market themselves as cultural capitals. Nevertheless, the gentrification of historic neighborhoods has resulted in sites of heritage consumption, cultural tourism, and displacement (Zukin, 1987; Jager, 1986: 78-91; Urry, 1990).[4]

From 2003 onwards, "Japanese Houses" became a valid category in historic preservation, largely resulting from community reaction against the National Taiwan University (NTU) plan to redevelop its old faculty housing around the main campus. The neighbors living nearby and some NTU students opposed the plan on varied grounds. Among them, some seniors were concerned with the old trees planted in the gardens of those houses while others attended to the architectural value and colonial histories reflected by the houses. A few of them paid attention to the underlying issue of privatizing state-owned lands, which urged NTU to redevelop those properties which otherwise would be acquired by the state. Therefore, NTU made a plan to redevelop the area comprising more than 33,000 square meters into high-rise research institutes and housing for visiting scholars. In response to the opposition, NTU eventually commissioned a research project to evaluate those dormitories built during the 1920s to '30s. To house its faculty, who were considered equivalent to governmental officials then, NTU built more than 150 houses around the campus. Since then NTU had been the major landlord in the city center over the century. The research team completed a thorough survey of the listed properties (Liu, 2003b). It suggested that those Japan-style houses collectively represented a colonial project of building a campus that ensured intellectual support for the imperial expansion in Southeast Asia. Moreover, many houses have been inhabited by important scholars over time, including Japanese scholars during the colonial era and Chinese scholars after 1949. These houses, along with their lush gardens, are significant cultural and ecological assets to the public. Eventually, NTU agreed to save some houses from demolition. It was from this case that the concept of "Japanese houses" became a generic term in urban heritage. Community movements, concerned for the protection of Japanese houses, soon came into being throughout the city, such as Cidong Street Neighborhood Association, which in 2004 was successful in countering Taiwan Bank and saving twelve Japanese houses.[5]

Chang Wei-hsiu, a researcher and participant in the cultural preservation activism, stresses the positive effect of "re-public-ization" of private dwellings through preserving urban heritage (Chen, 2012). In the study of the Qingtian Street conservation movement, Tso also argues that community effort to protect historical buildings is itself a Lefebvrian production of space (Tso, 2006). The processes of producing historical buildings shape the engaged citizens, contributing to a reclamation of the commons in the city. Yet the theoretical conceptualization of "restored common" is not necessarily accessible to everyone. NTU rented out the houses to a private party—that is, to whoever could restore the building and afford the rent. There is no clear guideline that requires the tenant to reuse those houses in line with the community's expectation. For example, a beloved Japanese house, which used to be home to a geologist/retired NTU professor, is now a high-end restaurant. The imagined common, as Tso notes, or the process to "re-public-ize the private" as Chang argues, is not stabilized or ensured in the subsequent adaptive reuse of those houses. Actually, the Qingtian Community Association was upset and felt dispossessed when they found themselves being implicitly excluded by the rent-seeking heritage management. Some of them argued that heritage preservation should have paid more attention to the legacies of the previous tenants. The case reveals clashes between privatization and building the commons; it also illuminates the very different approaches to heritage between the preservationists and the businessmen.

It is worth noting that the rediscovery of Japanese houses does not reveal much about how the presence of Japanese houses marked the colonial governance, a combined process of industrialization and urbanization. With construction of large-scale Japanese houses came the establishment of the Urban Planning Committee in Taiwan for the first time in history (1895). The city of Taihoku—the name of Taipei during the colonial period—had expanded each time the committee released an updated plan. The colonial government built staff housing around institutions or factories. Some houses were built by co-ops composed of Japanese homeowners. Today, about two thousand Japanese houses are still evident in the Daan, Chung-cheng, and Chung-shan districts in Taipei.

This kind of government-led housing construction stopped abruptly after 1945. From 1945 to 1975 there were only a few housing projects funded by the nationalist government (Chen and Li, 2010: 105-31). Only government officers, professors, and high-ranking army personnel would be housed in the left-behind Japanese houses. The grassroots immigrants mostly settled into self-built housing, which led to the massive construction of informal settlements. With the dream of returning to mainland China unlikely to be realized anytime soon, from 1956 onwards the army started to build low-rise collective housing near the Japanese houses to solve the desperate housing problem. For decades the expansion of existing Japanese houses into "Military Dependents' Villages" embodied the ambiguous welfare regime between the nationalist government and the mainland migrants in the period from the 1950s to 1980s. These Military Dependents' Villages often covered or absorbed existing Japanese houses. To some degree, this changing landscape reflects how the colonial hierarchy between the colonial and the colonized was replaced by class and ethnic differences between the mainlanders and the indigenous people.[6]

Beyond preserving the buildings, there is a hidden agenda of countering the strong tendency towards privatizing state-owned lands, a symptom of neoliberalization of the state in Taiwan since the 1990s. The neoliberal movement has been strengthened after the National Properties Management Committee and the "program for improved management of unused, underused, or occupied state-owned lands" was established in 2002. The National Property Bureau (NPB), which is the administrative arm of the committee, conducted audits of 10,701 state-owned houses, targeting the unused or under-used (under 50-percent occupancy) units that allegedly amounted to 76 percent of the whole. In this regard, institutions such as NTU, the Bank of Taiwan, the Taiwan Sugar Corporation, the Taiwan Railway Administration, and others were required to propose management plans within a given timeframe, otherwise NPB would resume the unused properties where many Japanese houses were located. NPB's goal is to revitalize properties—mostly by selling or renting to create public revenue. In other words, NPB is facilitating a process of privatizing public assets, which coincidentally led to a rediscovery of Japanese houses in the twenty-first century. In a way, it represented a break-up of the ambiguous welfare regime that exclusively benefited the mainland migrants.

In the next section I turn to the colonial aesthetics embodied in preserving Japanese houses and the rise of the creative city regime.

Colonial Aesthetics and the Creativity Regime: Fetishizing Japanese Houses and Colonial Aesthetics

Old trees and black roof tiles provide a quiet and shady refuge from the hubbub of the city. Chuang Kung-ju, in "A Green Legacy - The Japanese Houses of Taipei" (Tsai, 2005: 70)

This section critically examines the colonial aesthetics embodied in preserving Japanese houses and how emerging creative industries appropriate the constructs to build up a creativity regime. First, I discuss "colonial aesthetics" emerging from a fetishism of Japanese houses. Secondly, I will discuss the making of "creativity quarters" and the regime of the creative city. In recontextualizing the Japanese houses in the colonial city, I engage postcolonial theories as analytics to illuminate the ambiguous coloniality of Japanese houses reflecting the unusual colonial histories that condition urban thinking in contemporary Taiwan.

Paradoxically, the growing interest in rediscovering Japanese houses in Taiwan has alienated them from colonial history. Through the lens that focuses on materials, texture, and famous subjects who once lived in the houses, they were singled out from the larger socio-cultural and urban context.

Some architectural features of the Japanese house are highlighted, including the black ceramic roof tile, the decorative elements that marked the ridge, the elevated floor, and the wooden frames that composed the interior, and sometimes the Engawa.[7] The Japanese houses exhibited a simpler but delicate dwelling that manifests "Japanese" aesthetics.

Figure 2: The decorative tile that closes the roof ridge and the black ceramic roof tiles (courtesy of Chun-horn Cheng, 2013)

Representations of the Japanese houses often allude to naturalism in respect of "the green legacy" around the houses. Indeed, the ecological perspective contributes to the validation of preserving Japanese houses. With accumulating appreciation of their materiality, however, there is a tendency to fetishize and decontextualize Japanese houses.

Over the past decade, quite a few Japanese houses have become popular cafés and restaurants throughout Taiwan. A few became galleries, teahouses, or hostels. They are commonly seen in music videos, dramas, and films. Two decades before the fad, some famous directors filmed in the Japanese houses because they themselves grew up there, such as the Taiwanese director Hou Hsiao Hsien, with his autobiographical "A Time To Live And A Time To Die" in 1985. Today's media productions are less about biography than what I would like to suggest is "colonial aesthetics," the aesthetization of colonial effects. Today, actors creatively appropriate the colonial and meanwhile are colonized in fetishizing particular materiality and sensibility of the colonial past without necessarily reflecting on the historical formation of colonialism. In the particular case of rediscovering Japanese houses in Taiwan, it is a combination of colonial and anti-colonial reaction to validate "the colonialized," but the role of the colonizer is unstable. In practice, the active cultural agents often elaborate on "aesthetics of Japanese houses" rather than "colonial aesthetics." It is the rhetorical choice, I argue, that sustains the power of the colonial.

It is worth noting whether restoration and representation of Japanese houses can invoke postcolonial sensibilities. In the list of film settings recommended by the Taipei Film Commission lies the category of "Japanese Architecture," which included some major institutional buildings and six quarters composed of Japanese houses, including Cidong Street and the Wistaria Teahouse.[8] Outside Taipei there are also quite a few places actively advertised as suitable film settings, such as the staff housing of the Taiyang Mining Company in Jingtong. These representations of Japanese houses present a manufactured nostalgia without clear subjects. They do not necessarily include those who inhabited the houses under Japanese occupation or the nationalist government of 1950 to 1990 unless they are well-known to the public. They rarely present the postcolonial sensibility of these dwellings, nor do they invite the audience sitting in the theater to become "transient inhabitants"[9] of postcolonial sensibilities. Historical details are certainly not as important as the architectural details in the wedding photography sited against those houses. Actually, those decorative elements had been popular in the antique market. The demand rises from antique collection and the growing need to renovate abandoned Japanese houses. The tiles are particularly expensive because they are not produced in Taiwan anymore; there used to be plants built in colonial times that served as suppliers for the massive construction throughout the island. Considerable looting of vacant houses occurs in search for building materials; however, no serious action has been taken to address this issue.

Japanese houses were integral to a broader project of industrializing Taiwan to serve Imperial Japan. In a sense, the colonial government was one among many major constructors in Taiwan. Other colonial enterprises in mining, sugar, lumber, railways, wine, and institutions like universities and schools, prisons, and armies all participated in the massive construction of houses to accommodate the extensive bureaucracy. Since the 1930s, private investment in housing construction grew to accommodate the rapid increase of Taiwan's population. In the period 1895 to 1945, the population rose from nearly three million to six million. Yet, whether in local academia or preservation practices, Japanese houses are often singled out from the larger context. Shin-An Chen's dissertation is the first treatment that gives an overview of the institution of construction (2004). Most research focused on a particular building or only one residential quarter. In the important cases of the Sungshan Tobacco Factory Park and the Huashan Winery Park, the residential neighborhoods built next to the parks formed a self-contained settlement (some even like a small town), and comprised schools, hospitals, libraries, baths, and club houses. The staff who worked for the Sungshan Tobacco Factory during the 1920s to 1930s were equivalent to today's middle-class in the city. These housing sectors also testified to a pre-war welfare regime throughout the industrialization process, which was sustained until deindustrialization and neoliberalization started in the late 1980s. Yet, the historical context reflected in the urban fabric was fragmented and dislocated in the practice of historic preservation. While the Sungshan Tobacco Factory was turned into a park for the creative industry (a so-called "creative park"), the residential quarter was demolished. As some houses associated with these industrial enterprises became popular cases of adaptive reuse, they were decontextualized and treated as stand-alone objects of fetishism.

There is no room in this essay for a comprehensive discussion of colonial housing construction and renovation. However, I would like to stress the particularly ironic history of mainlanders' relocating to Japanese houses in Taiwan. As mentioned, the left-behind Japanese houses were used as ad-hoc housing to accommodate the influx of mainlander migrants settling in Taiwan after 1949. More than one million people arrived in Taiwan and immediately caused a housing crisis (Spear, et al., 1988). Most migrants considered the move to be temporary, which therefore rendered living in the homes of their enemies less uncomfortable. Looking forward to a return to their homelands, everything exceptional can be justified by the belief that the island named Taiwan was just a transitory stop. In this light, high-ranking generals along with lieutenants moved into those Japanese houses and started to experience cultural differences embodied in their spatial organization and interior design. Generals of the air force lived in houses on Cidong Street, and professors of NTU (originally Taihoku Imperial University) lived on Qitian Street. Many of them did not live to return to the mainland; the ban on visits between the two sides across the strait was not lifted until 1987. Setting aside the emotions, it was challenging for the majority to get used to the "floor living" of sitting and sleeping on the tatami. The habit of taking off shoes inside the house was also foreign to most mainlanders. When realizing that the relocation might be an extended stay—one without end—they started to change the flooring and add furniture to reconstruct a more familiar setting in their "Japanese houses." The aforementioned irony and difficulty of living during the years from 1949 to the late 1990s, however, is largely ignored when featuring Japanese houses. In some cases, the post-war history is ignored and reversed—by restoring the houses back to how they should look—to highlight the authenticity of Japanese houses.

It would seem unthinkable for postcolonial scholars to see that the rediscovering of Japanese houses in Taiwan has not been examined in a context of (post) colonial cities, which may itself epitomize the issue of contemporary urban studies in Taiwan as a product of colonialism. That the colonial city and its history ended in 1945 is taken for granted, as *a priori*. Colonialism, as Jane M. Jacobs notes, "entails the establishment and maintenance of domination over a separate group of people, who are viewed as subordinate, and their territories, which are presumed to be available for exploitation" (1996: 16). Colonialism is a particular articulation of imperialism (Said, 1979). Let me keep the definition at the outset and then borrow Shu-mei Shih's critiques of postcolonial theories (2001: 709-18). In developing the concept of Sinophone, Shih points out that postcolonial theories, largely developed by the diasporic scholarship in the West, are limited by nationalist historiography. Shih argues that the flip side of an unreflective nationalism can be a new imperialism. In the case of China, most historical analysis ignored the Qing's territorial expansion since the eighteenth century, which qualified itself as a "continental colonialism" (2001 and 2007). If we take Shih's idea to rework our understanding of the 1945 takeover of Taiwan, the 1949 migration, and the continuous migration that can actually be traced back to the seventeenth century, it would seem that there is more than one colonialism at work. If we explore this experimental idea in cultural terms, the history of mainland migrants settling in Japanese houses in Taiwan indeed embodied a kind of in-between colonialism under the "black tile rooftop," the symbolic element of the commodifiable colonial aesthetics, or in other words, a decontexualized aesthetization of oppressive colonial history.

My critique of colonial aesthetics associated with Japanese houses is looking beyond the commodification of colonial memories in Taipei to interrogate the materiality of imperialism, as Anthony King suggests (2003: 167-86). King draws attention to "material practices and real politics" of the "actually existing postcolonial urbanism and architecture." The Japanese house deserves a postcolonial analysis of its impact of shaping life of the colonial and the colonized long after the Japanese left, as the landscape historian Amita Sinha does with the bungalows in Lucknow (1999: 56-63). The inhabitation/colonization carries on with the mainlanders moving in as new tenants. The mainlander migrants, who were not the colonized, have been living in the postcolonial dwellings left by the colonizers. The indigenous, instead, were those who watched the hectic transition as the gaze of the historical landscape has been gradually layered by conceptualization of "the mainlander class." The colonial hierarchy was soon replaced with new class relations, which left the colonized no time to reflect on the colonial past. In the early years following 1949, only some indigenous people visited these places (sites of memories) if they had developed a relationship with the mainlander class; for example, those who worked as maids to serve the mainlander families. Sometimes the indigenous children were invited by their classmates, who were born to those mainlander families. Later on, the social interaction became more hybrid when inter-group marriage started in the military dependents' villages. These hybrid encounters contribute to the formation of hybridity in multiple sites, such as the kitchens where the indigenous women learned to prepare distinctive cuisine from northern China for their employers or husbands. As King notes, "Multiple identities produce a multiplicity of spatialities" (2003: 172). The same is true in reverse. The different regimes embedded in multiple spatialities can trap inhabitants in different temporalities and geographies. Mainland migrants living in the Japanese houses, in a way, were trapped in the (post) colonial condition, which is co-produced by the colonial materiality of the houses and the (post) colonial anxiety experienced by the indigenous. Within the Japanese houses, the most intimate settings for the residents, the replacement of colonialism indeed cultivates a (post) colonial sensibility mixed with continuation of colonialism that deserves more attention before the houses are remade into cafés and restaurants selling British tea or American steaks.

There has been little critical reflection of the postcolonial sensibility with regard to planning the physical environment in Taipei. The rapid change of Taipei's urban landscape and the selective preservation of Japanese houses is not only a loss of "lieux de mémoire," environments of memory, but also a missed opportunity to trace postcolonialism in "lieux de mémoire," the site of memory (Nora 1989: 7-24). In forgetting colonial memories, we see a repackaging and aesthetization of memories around heritage to promote tourism and the creative industry. The "fascination with heritage" to re-inscribe the postcolonial landscape, if seen as a key mechanism in defining community or national identity, leads to a confusing conceptualization of Japanese houses in the past decade. Perhaps they epitomize Homi Bhabha's notion of "hybridity" (1994), which simultaneously returns agency to the colonialized and contributes to sustaining colonial power and sensibility. Ideally, postcolonial actions can be a creative remaking of the colonial past by the colonialized. But in the case of Japanese houses in Taipei, the colliding colonialisms at the same place do not comply with a binary concept of the colonial and the colonialized. It exposes the limit of theories and how important it is to understand formations of postcolonialism in the material politics of everyday life rather than linear time.

The Making of "Creative Quarters" and the Creative City Regime

Heritage is not in any simple sense the reproduction and imposition of dominant values. It is a dynamic process of creation in which a multiplicity of pasts jostle for the present purpose of being sanctified as heritage. (Jacobs, 1996: 35)

In his recent work, Michael Keith discusses the tendency to capitalize on cultural hybridity through creating "creative quarters" in the cities, which can lead to "the conceptual overexploitation of the notion of the hybrid as simultaneously commodity, ethical stance, and measurable aesthetic" (2005: 132). These "creativity quarters," as metonymic of the city and as "an object of government," are considered to be indicative of advancement and success (2005: 128). Through processes of selective celebration of diversity and ethnicity, dilapidated ethnic enclaves are turned into vibrant, artistic creative quarters, which reminds us of the kind of "artistic mode of production" that Sharon Zukin discusses in her study of New York (1988: 176-82). The development of creative quarters is largely an artistic redevelopment strategy that creates historic preservation.

Cultural tourism is one of the new economies enabled by, and used to justify the making of creative quarters. Growing tourism around heritage in Taipei, Singapore, and many other cities seems to testify to the power of culture and creativity. The pundit of creative cities Charles Landry[10] once lamented Singapore's loss of cultural heritage. To challenge the stereotype that Singapore cannot be a cultural capital, cultural heritage became a focus of urban policy, leading to the remaking of China Town, Little India, and Kampong Glam to make Singapore "A Magic Place of Many Worlds."[11] Ethnic difference was rediscovered and shophouses in different ethnic quarters became assets that can attract foreigners who crave "the Orient." In addition, preservation was seen as an effective way to foster identity and bind people together in Singapore (Yeoh and Kong, 1997: 52-65). There is certainly a cost in identity-making around heritage. Along with upgrading the built environment and commodification of ethnic culture comes obliteration of inconvenient memories and displacement. The beautiful veneer imposed on the heritage of Kampong Glam hides the eviction of the school that played an important role in Islamic education, and the descendants of Sultan Hussein Shah from the family's ancestral home in the palace at the heart of the Kampong Glam. The ancestral home later became the Malay Heritage Center to showcase Malay culture. The reified image of ethnic culture displaced the early tenant's everyday life, and there is always a social price.

A similar issue of commodification of ethnicity is also present in Taipei. There were historical memories of displacement, including those of Japanese repatriates, Taiwan-born Japanese, and the mainland migrants. There is massive displacement as an immediate outcome of heritage preservation and urban regeneration, which is largely ignored in marketing those heritage-turned-creative quarters. Yet the approach is encouraged by city policy makers and welcomed by consumers and tourists. Following Singapore, the Taipei city government invited Landry to the city several times to advise on how to improve the "creative milieu" of Taipei.[12] Landry highlights the analogy between the form of creative network and the built form of historic quarters, suggesting that the density and diversity comprised by old buildings and narrow, winding streets contributed to a more interesting built environment to nurture social interaction and cooperation. Adaptive reuse of historical buildings, just like Japanese houses, are validated in his assessment and repositioned as nodes in making creativity hubs.

Indeed, reutilizing the industrial heritage into creative quarters becomes a normalized approach in Taiwan. Among them, "Huashan 1914" is the most notable example. The Huashan winery was built by the colonials in Taipei in 1914. It was shut down in 1987 and had been abandoned for almost a decade before it was rediscovered by a group of artists. Thanks to the saga of saving the Huashan winery from redevelopment from the mid-1990s to 2002, and to the growing attention given to cultural and creative economies at that time, the Taiwanese government launched the plan of developing five Creative and Cultural Parks across five cities in 2002, a result of a "creative industry" that was becoming an important agenda at the national level for the first time in the same year. In this light, the Huashan winery has been rebranded as "Huashan 1914," a creativity hub where creative events and performances take place. Some considered the case as indicative of a paradigmatic shift from heritage preservation toward a creative economy (Cheng and Fu, 2011); some considered it as failing to go beyond the regeneration of the park (Wang, 2011). Some foreign scholars, from the perspective of an international tourist, appreciate its interesting, urban ambience (Fernández-Muñoz and Hergueta, 2012). In general, the consumption-oriented creative economy of the park has received more critiques than compliments. Today, twelve restaurants, four design shops, and retail businesses including books, music, yoga dress, and so forth, occupy the 7.2-hectare park. Recently, a movie theater moved in to one of the old wineries. It has become the most popular entertainment quarter for night-life rather than a creative quarter. Like frontier gentrifiers in the scholarship of gentrification, the artists who participated in preservation activism were all displaced.[13] Not much attention is given to the wineries and the urbanization processes along the railway. The regeneration of the winery park is largely separated from the urban context, in which Cidong Street plays an important role.

As the culinary business dominates the creative park, there is a sense of cosmopolitanness exhibited "on the table," which, however, replaces the multiple historical passages that once intersected in the wineries and Japanese houses. A variety of cuisine styles, including American burgers, Irish bars, pizzerias, and Japanese sushi, turn the park into a popular dining place. Some wineries are open for concerts and private parties. It demonstrates the generalization of reducing the creative industry into making "creativity quarters," which, in many cases, were just a repackaging of the consumption landscape. It is apparently far from the kinds of creative industries characterized by the generation and exploitation of intellectual property as adopted first in the UK context and elsewhere,[14] the rise of the creative class (Florida, 2002), or the intense recursive relations between place and the logic of local production systems of cultural goods.[15] "Creativity" seems to be a theme for marketing rather than the substantive of economies in the case of Taipei.

The Creative City Regime

The Sungshang Tobacco Creativity Park (STCP) is the most recent example of making a creative quarter in the city of Taipei. In September 2013, there was a pretentious exhibition set in STCP, which featured Taipei's potential to become the "World Design City (WDC)." The exhibition was part of the city's effort to win the bid for the 2016 WDC, an inter-city competition that capitalizes on culture, art, and creativity. Actually, "the Creative Cities Network" promoted by UNESCO since 2004 is the more prominent one that features symbolic economy driven by creative industry. The discourse of making creative cities has renewed its languages in strategizing urban development and economies. Charles Landry, the well-known advocate for creative cities, traveled around the world to discuss with city mayors strategies to make their cities more creative. In celebrating creative economies and innovative partnerships between the public and the private, the discourse, policies, and stakeholders who foster the policy transfer and inter-city exchange over "creative cities" to attract investment and encourage production of cultural goods constitute the form of governance that I call "the creative city regime." It is such a regime that justifies construction of creative quarters and encourages colonial aesthetics.

The celebration of creative industries, which is largely based on developing the knowledge-based economy, is translated into a discourse of the creative city that highlights physically constructing creative spaces and industry clusters rather than fostering cultural capital and infrastructure in contemporary cities (Scott, 1997; Comunian, 2011: 1157-79). Creative economies cannot come into being without involvement of technologies, innovative organization of production, and creative labor, all of which are conditioned by the dynamics of locational agglomeration. The creative economy rises from the "pooling effect" and "clustering." It has to go beyond commercialization of heritage and urban regeneration; it is the ever-changing articulation between capitalist production, culture, and (re)formation of place (Scott, 2010: 115-30, and 2006: 1-17). Yet, as Comunian points out, much of the focus has been around investing in a specific type of urban regeneration project or flagship developments rather than addressing the nature of the infrastructure, networks, and agents engaging in the city's cultural production and consumption. In many cases, the discourse is used to justify a wide range of redevelopment in inner-city neighborhoods and the former and residual industrial landscape at the waterfront. Cities around the world utilize "the creative quarter" as a panacea to implement broader city expansion and regeneration plans (Evans, 2009: 1003-40). The creative city concept has served the goal of a specific set of stakeholders to enhance the identity of the city and foster the real estate market and businesses but not necessarily creative entrepreneurship. Cases can be seen in the Entertainment District in Toronto, creative industry clusters in Shanghai, or art-led revitalization in Wedding, Berlin.[16] Though much has been said about the importance of micro-interactions and networks between creative practitioners, investment often goes to enhance the creative outlook of the city instead, the so-called "culture-led regeneration." Among many, Bilbao and Glasgow are two prominent "cultural capitals," which continuously capitalize on culture and creativity (Johnson, 2009: 6).

As much as neoliberalism controls cities through "freedom"—market freedom—the creative city regime exercises its power through "creativity," a new name for post-modern consumption culture. As Zukin suggests, "Culture is a powerful means of controlling cities. As a source of images and memories, it symbolizes 'who belongs' in specific places" (1995: 1). In the name of creative cities, culture-led urban regeneration certainly leads to exclusion and displacement of those who are non-creative. In the case of the Huashan 1914, the creative city regime displaces everything but creativity. But "what is creativity" is never openly negotiated, although exhibited in shops, menus, and tickets to performances. The commodified diversity on the table is not even the "staged authenticity" (MacCannell, 1973: 589-603). It can be totally disconnected from the multiple pasts. There is no attempt to negotiate the ideological, social and material structures of power established under colonialism during the postcolonial period. Perhaps there is no postcolonialism yet (Jacobs, 1996: 25).

Making a Creative Quarter against the Prison Wall of the Taipei Prison Settlement

In October 2013, the central government announced a renewed plan to redevelop the Huaguang neighborhood in the center of Taipei into a Roppongi Hills–like business district and a creative quarter composed of twenty-two Japanese houses and a beautiful red-brick public bath on Jinghua Street. The historical houses would be restored to accommodate activities that support young people's participation in creative industries. No clear reference was given for the public to understand the connection between Roppongi Hills (a model borrowed from Tokyo), the place named "Huaguang," and the suddenly imposed creative industry. There is only a brief mention of "the prison wall" in the official announcement, which implicitly points to the forgotten colonial prison settlement in the relics.

Figure 3: The Taipei Prison Settlement at the Huagaung neighborhood
(courtesy of Lian-sheng Cheng, 2013)

More than a century ago, the area where the Chiang Kai-shek Memorial Hall and Huaguang now stand was a vast field outside the city. It was not urbanized until the Japanese colonizers arrived in 1895, making an ambitious decision to construct a grand, modern prison even before constructing the Governor's House. Interestingly, the British colonizers took the same initiative in Hong Kong (then Victoria City) at the end of the nineteenth century. As geographer Gilmore notes, the construction of prisons is a "project of state building" (2002: 16). In Taipei, the footprint of the prison in the then southeastern suburb clearly manifested the colonial presence by simultaneously demolishing the old city walls of Taipei built by the Qing dynasty and constructing Taihoku Prison (Taihoku means Taipei in Japanese). The 900-meter-long old city wall was broken down and became the four-meter-high walls around the prison in 1904.[17] Even the Japanese at home were amazed by the sheer size and architectural design of Taihoku Prison. The colonizers were very proud of constructing the most beautiful sight in Taipei (Botsman, 2005: 209). With the European idea of punishment/correction by detention, the colonial government laid out the whole new correctional institution composed of thirteen prisons in Taiwan. As the most important correctional project, the

Taihoku prison had an orderly footprint that marked the accomplishment of keeping the colony under control and turning it into a civilized city.

Taihoku prison, with its radial floor plan and system of solitary confinement, exemplifies the idea of the panopticon as designed by Jeremy Bentham[18] and the Pennsylvanian prisons that emphasize self reflection (Teeters and Shearer, 1957). Incarcerated there, prisoners were made to work in factories and farms, and it was believed that "the labour by which the convict contributes to his own needs turns the thief into a docile worker" (Foucault, 1979; 243). The prison was constructed as a self-contained settlement, which comprised the penitentiary; hundreds of official residences and dormitories to accommodate the officials and staff of the prison; a training place for the prison staff to practice Kendo and Judo; and amenities such as wells, baths, a farm, and even a graveyard. Within the walled prison there were cells, factories, hospitals, and other infrastructure. By the end of the Second World War, the whole prison settlement accommodated thousands of people outside of the city.

Displacement, and Privatization of State

The postwar Taipei prison became located in the center of Taipei as the city grew rapidly during the 1950s and 1960s. Thus, the government decided to relocate the prison outside the city. The prison building was demolished and the land was sold to two state-owned enterprises, Chinese Telecom and Chinese Post Office. The two complexes stood oddly next to the prison walls and the historical settlement, which continued to accommodate the prison warders and officers. Postwar migrants managed to build shacks in-between Japanese houses during the housing shortage. The informal construction was accepted in a time of transition, a prolonged period that nobody could imagine at that time.

Though haunted by the fearful memories of prison, the prison settlement is a relatively affordable place for the underprivileged in Taipei because of lack of investment. Until 2007, it had become a neighborhood named Huaguang, inhabited by more than three thousand people. National flags were hung by the veterans below the Japanese black roof tiles; palm trees planted by the Japanese stood next to the Chinese toon later added by the mainland migrants.

The layers of landscape reflected the complex place memories of Huaguang. The former colonial prison settlement gradually transformed into a hybrid settlement on the state-owned land.

Given its prime location, the central government has planned to redevelop it since the early 1990s. Several ambitious proposals were delivered, but nothing came of them. Among many, the most recent plan, released in 2011, is to construct "Roppongi Hills in Taipei" (RHT). It is a project that aims to redevelop the place into high-end shopping centers and luxurious hotels by selling development rights to private developers. The RHT project has never gone through public consultation. The state-owned project bypassed citizen participation as required by the city government. Despite criticisms, it displaced all of the residents and left only one temple, whose followers still offer strong resistance. There was no compensation or rehousing. About two hundred families even found themselves being sued by the government on the charge of illegal occupation and "unjust enrichment."[19] The charge of illegality deliberately ignores the history of the settlement. The displacement of people and histories indeed facilitated the government's interest in harvesting the handsome land rent. It was not until the spring of 2013 that several civic groups formed an alliance, calling for the preservation of the colonial prison settlement and old trees (Ho, 2013). A small success came after six months of battle. Twenty-two Japanese houses and one public bath were designated as historic buildings out of the more than 150 houses. The central government was forced to revise the RHT project, downsizing the development, designating a part of the project area as a heritage district for creative industries.

Decontextualized Heritage Taken Over by the Creative Regime

The reason why the creative industry is suddenly included in the project is worth noting. On the one hand, the adoption of the creative industry and programs for creating jobs for young people is seemingly a gesture responding to the alliance's criticism that the RTH project pays no attention to city culture and livelihood. On the other hand, it demonstrates that the creative industry has become a convenient strategy to address the combined issues of representation of culture, heritage management, and community development. Yet, with the selected preservation of heritage and with top-down planning, it is doubtful whether the Japanese houses taken out of the context could be turned into an effective creative quarter where history and innovation illuminate one another.

Historic preservation is inevitably about remaking geographies of colonialism in most postcolonial nations. Yet, the story of the construction and demolition of the colonial Taihoku Prison involves multiple colonizations, oppressions, and displacements across three centuries. The fact of a decontextualized preservation of Japanese houses right next to an ambitious plan to replicate the Japanese model of Roppongi Hills is ironically engaging "the neocolonial" to replace the colonial prison and the informal settlement in-between nations and memories. It is not that easy to identify the colonial and the "post." What is clear is that the unfortunate displacement and charges of unjust enrichment somehow resonate with the punitive landscape where the colonial state exhibits its power over the citizens.[20] Meanwhile, the creativity regime is taking over the contested terrain where citizens and communities still cry out for alternatives—genuine creative solutions rather than the same old wine in a new bottle. Whether or not the RHT will be realized remains a question at the time of writing. The decision over what kind of creative quarter will be orchestrated with the future restoration of the Japanese houses on the edge of the site against the prison wall is still in a state of flux.

Conclusion: Calling for a More Creative Way to Restore Postcolonial Sensibilities

The photography collection titled "Japanese House" by Tomoko Yoneda, as mentioned in the beginning of the chapter, has traveled across countries, sharing with the international audience the photographer's rediscovering of the colonial footprints of Imperial Japan. In the meantime, Taipei—the city where Yoneda took those photos—is promoting itself as the wouldbe WDC. Part of its effort is seen in the way in which the construct of "Japanese houses" takes on new lives and new meanings in contributing to the image of the creative, adaptive city in the past decade. The creative appropriation of the colonial past interestingly presents the city as a unique postcolonial project, which is slightly different from Jacobs' argument that "the move to formal independence is shot through imperialism itself" (1996: 16). The case of absorbing Japanese houses into part of the creative city appears as a Taipei-style cosmopolitanness that neutralizes if not counters imperialism. It presents the cosmic liberation of the Japanese houses from the burden of histories and ethnic conflicts. Yet, as I pointed out earlier, the replacement of colonial histories with creative practices also risks displacing the inhabitants of postcolonial sensibilities, whose experiences are crucial for us to reflect critically on the oppression and exploitation of colonialism. What is worse, the so-called making of creative quarters can be anything but genuinely developing creative economies. What is at work is predominantly "the myth of new urban culture" promoted and dominated by "the tourism coalition" (Hoffman, et al., 2003). The creativity regime in Taipei, I am afraid, is just repackaging similar strategies with new languages, though it may be too early to conclude that at the moment.

My goal in writing this chapter is not to argue for a correct way of telling stories about the Japanese house as urban heritage. What fascinates me and others who pay attention to politics of colonial heritage is the confusing representation of postcolonialism that cannot be easily identified and located but has been conveniently commodified. With the ambiguous postwar years adding another layer of "in-betweenness" onto the built environment, the colliding colonialisms found within the Japanese houses in Taipei do not comply with a binary concept of the colonial and the colonialized. The photographer captured the postcolonial quality in her photo that presents the wallpaper and posters added by the mainlander migrants;[21] the consumers and the tourists who visited those houses partook in a fetishization of colonial aesthetics. Rising from these practices is a neocolonial formation of the creative city regime, which objectifies "the Japanese house" within a broader colonial context of capitalist industrialization. As discussed in the case studies, the historic preservation activism interrogates the privatization of state and urbanization of neoliberalism in saving some Japanese houses from demolition. Unfortunately, the heritage-turned-creative quarter, as a collection of Japanese houses or factories, is soon taken over by the creative city regime. In turning colonial history into a decorative element like the ceramic tile, the regime successfully displaces the historical housing issue and expropriates the lands, reinforcing "the globalization of colonial modernity," as Arif Dirlik points out, in the guise of making creative cities (2007). Gazing at the Japanese house in Yoneda's work, I cannot help but wonder if the people who step into those houses against the prison walls can find a more creative way to restore postcolonial sensibilities that will reveal the colonial present.

Notes

1. Another way to put it is to categorize those houses as "Japanese-style-dormitories" (literally meaning 日式宿舍 in Chinese). But the term is a bit misleading as not necessarily every Japanese-style house was built by the colonial government.
2. "Unraveling history in 'Japanese House' – Tomoko Yoneda at ShugoArts." http://artradarjournal.com/2011/11/16/unravelling-history-in-japanese-house-tomoko-yoneda-at-shugoarts.
3. According to the Cultural Heritage Preservation Act in Taiwan, heritage listed as "monument" is of the most significant value, and its authenticity should be restored and protected. "Historical building" refers to secondary heritage that can be adaptively reused.
4. See a thorough discussion of cultural tourism in Urry, 1990.
5. Nine out of the twelve houses are located on Cidong Street while the others stand within the same street block.
6. Here "the indigenous" is defined in a broader sense to include the Taiwan-born Hans, whose ancestors arrived in Taiwan between the seventeenth and twentieth centuries. When used in its strict sense, "the indigenous" refers to the aborigines who had been living on the islands before major Han Chinese immigration. Culturally and linguistically, they belong to the Austronesian group.
7. *Engawa* refers to the typically wooden strip of flooring that immediately connects with the windows and storm shutters inside traditional Japanese rooms. Sometimes it has disappeared under major additions of extra rooms built after the mainlander inhabitants moved in.
8. The list is available at http://www.taipeifilmcommission.org/tw/LoveMovie/LocationList/1705.
9. See an elaboration on the idea in Yeo, 2003: 258-61.
10. Charles Landry is the author of the book *The Creative City: A Toolkit for Urban Innovators* (2008).

11. The idea was adopted in the 1986 Tourism Product Plan; see Johnson, 2009: 169; and Henderson, 2004: 113-25.
12. After the visit, the Urban Regeneration office published a bilingual booklet that presents Landry's assessment (Landry, 2012).
13. In saving the Huashan Winery, a group of artists formed the Association of Culture Environment Reform Taiwan (ACERT) in 1998. Until the winery was transformed into its latest version (Huashan 1914), ACERT had been the most important player in overseeing the restoration of the Huashan Creative Park, but it had been marginalized if not displaced from the park. Since 2007, the park has been operated by the Taiwan Cultural-Creative Development Company, a body co-founded by the Ambassador Hotel, a local publisher, and a design firm.
14. The UK Government Department for Culture, Media and Sport (DCMS) stresses "individual creativity, skill and talent" and the potential to develop wealth and job creation through "the generation and exploitation of intellectual property." The DCMS emphasis on intellectual property has been influential; see Department for Culture, Media and Sport 2001.
15. Scott, 1997: 323-39. In Scott's case study analysis of the film industry in Los Angeles, he stresses the symbiotic relationship between place, culture, and economy. He also pays attention to the dense human interrelationships that give rise to an innovative network of cooperation and transmission of information service. Also see Scott, 2005.
16. See the Toronto case in Darchen, 2013: 188-203; see the Berlin case in Jakob, 2013: 447-59; and see the Shanghai case in Zheng, 2012: 3561-82.
17. The Taipei Prison was characterized by the stylistic Embuko, which is not seen in Western prisons. See the special issue on the Taipei Prison, translated by H. Lin (2013), in *The Taipei Penal Institution Monthly*, The Taipei Penal Institution. (Original work only available in Japanese, published on 1 May 1938).
18. Jeremy Bentham noted his design of "panopticon" in *The Rationale of Punishment* (1830).

19. "Unjust enrichment" is a term used in the common-law system, meaning when a person unfairly benefits by chance, mistake, or another's misfortune, and the one enriched has not paid or worked for the benefit, morally and ethically they should not keep it. An obligation to make restitution arises, regardless of liability for wrongdoing.
20. Shu-Mei Huang, *Tracing the Punitive State: Punishment and Displacement in Remaking the City,* working paper.
21. For example, the one titled "Former house of General Wang Shu-Ming, the Chief of Staff under Chiang Kai-Shek, Cidong Street, VI," 2010.

Bibiography

Ashworth Gregory. "Paradigms and Paradoxes in Heritage as Development." Paper presented at Revitalising Built Environments: Requalifying Old Places for New Uses, Istanbul Technical University, 12-16 October 2009.

Bhabha, Homi K. *The Location of Culture*. London: Routledge, 1994.

Botsman, Daniel. *Punishment and Power in the Making of Modern Japan*. Princeton: Princeton University Press, 2005.

Chen, Hsin-yi. "Bringing Old Japanese-Era Houses Back to Life." *Taiwan Panorama*, April 2012.

Chen, Shin-An. "The study on the 'Building standards of official residence' of the Taiwanese Governor General's Office during the Japanese Period." PhD diss., National Cheng Kung University, 2004.

Chen, Yiling, and William Der-Hsing Li. "Neoliberalization, State and Housing Market: Transformation of Public Housing Policies in Taiwan." *The Journal of Geography* 59 (2010): 105-31.Cheng, Min-Tsung, and Chao-Ching Fu. "Beyond Heritage, Towards the Possibility of Creative Economy: The Case of the Reuse of the Industrial Heritage in Taiwan." Paper delivered at ICOMOS, Paris, 2011.

Comunian, R. "Rethinking the Creative City: The Role of Complexity, Networks and Interactions in the Urban Creative Economy." *Urban Studies* 48, no. 6 (21 April 2011): 1157-79.

Darchen, Sébastien. "The Creative City and the Redevelopment of the Toronto Entertainment District: A BIA-Led Regeneration Process." *International Planning Studies* 18, no. 2 (March 2013): 188-203.

Department for Culture, Media and Sport. "Creative Industries Mapping Documents 2001." London, 2001.

Dirlik, Arif. *Global Modernity: Modernity in the Age of Global Capitalism.* Boulder, CO: Paradigm, 2007.

Evans, Greame "Creative cities, creative spaces and urban policy." *Urban Studies* 46, (May 2009): 1003-40.

Fernández-Muñoz, Laura, and Aurora Galán Hergueta. "Hybrid and creative Taipei." In *Ambiances in action. Proceedings of the 2nd International Congress on Ambiances/Ambiances en acte (s). Actes du 2nd Congrès International sur les Ambiances.* Montreal, 2012.

Florida, Richard. *The Rise of the Creative Class: And How It's Transforming Work, Leisure and Everyday Life.* New York: Basic Books, 2002.

Foucault, Michel. *Discipline and Punish: The Birth of the Prison.* Vintage Books, New York, 1979.

Gilmore, R.W. "Fatal couplings of power and difference: notes on racism and geography." *The Professional Geographer* 54, no. 1 (February 2002): 15-24.

Hoffman, Lily M., Susan S. Fainstein, and Dennis R. Judd. *Cities and Visitors: Regulating People, Markets, and City Space.* Malden, MA: Blackwell Publishers, 2003.

Henderson, Joan C. "British colonial heritage in Malaysia and Singapore." In *Tourism and Postcolonialism: Contested Discourses, Identities, and Representations*, edited by C. Michael Hall and Hazel Tucker, 113-25. London: Routledge, 2004.

Jacobs, Jane M. *Edge of Empire: Postcolonialism and the City.* London: Routledge, 1996.

Jager, Martin. "Class Definition and the Esthetics of Gentrification: Victoriana in Melbourne." In *Gentrification of the City*, edited by Neil Smith and P. Williams, 78-91. Boston: Allen & Unwin, 1986.

Jaivin, Linda. "Wistaria Teahouse." *China Heritage Quarterly* 29 (2012).

Jakob, Doreen. "The eventification of place: Urban development and experience consumption in Berlin and New York City." *European Urban and Regional Studies* 20, no. 44 (1 October 2013): 447-59.

Johnson, Louise C. *Cultural Capitals: Revaluing the Arts, Remaking Urban Spaces.* Farnham, UK: Ashgate, 2009.

Keith, Michael. *After the Cosmopolitan? Multicultural Cities and the Future of Racism*. London: Routledge, 2005.

King, Anthony. "Actually Existing Postcolonialisms." In *Postcolonial Urbanism: Southeast Asian Cities and Global Processes*, edited by Ryan Bishop, John Phillips, and Wei Wei Yeo, 167-86. New York and London: Routledge, 2003.

Landry, Charles. *The Creative City: A Toolkit for Urban Innovators*. New Stroud, UK: Comedia, 2008.

— "Talented Taipei and the Creative Imperative." Taipei City Urban Regeneration Office, 2012.

Liu, John K.C. "Study of the Colonial Buildings and Dorms Managed by National Taiwan University." Taipei: Graduate Institute Building and Planning, 2003a.

— "The Study of the Wistaria Teahouse." Taipei: Department of Cultural Affairs, Taipei City Government, 2003b.

MacCannell, Dean. "Staged Authenticity: Arrangements of Social Space in Tourist Settings." *American Journal of Sociology* 79, no. 3 (November 1973): 589-603.

Nora, Pierre. "Between Memory and History: Les Lieux de Mémoire." *Representations* 26 (1 April 1989): 7-24.

Scott, Allen J. "Creative Cities: Conceptual Issues and Policy Questions." *Journal of Urban Affairs* 28, no. 1 (January 2006): 1-17.

— "Cultural economy and the creative field of the city." *Geografiska Annaler. Series B, Human Geography* 92 (12 August 2010): 115-30.

— "The Cultural Economy of Cities." *International Journal of Urban and Regional Research* 21 (2 November 1997): 323-39.

— *On Hollywood: The Place, the Industry*. Princeton: Princeton University Press. 2005.

Shih, Shu-mei. "The Concept of Sinophone," *Modern Language Association* 126, no. 3 (2001): 709-18.

— *Visuality and Identity: Sinophone Articulations across the Pacific*, Berkeley: University of California Press, 2007.

Sinha, Amita. "The Bungalows of Lucknow," *Open House International* 24, no. 2 (1999): 56-63.

Speare, Alden, O-chin Liu, and Ching-Lung Tsay. *Urbanization and Development: The Rural-Urban Transition in Taiwan*. Boulder, CO: Westview Press, 1988).

Teeters, N.K., and J. D. Shearer. *The Prison at Philadelphia, Cherry Hill: The Separate System of Penal Discipline, 1829-1913*. Philadelphia: Temple University Press, 1957.

Tso, Hsiang-Chu. "Historic Building Conservation as Social Production of Space—The Movement of Japanese Style Building Conservation in Qingtian Street, Taipei." Master's thesis, Graduate Institute of Building and Planning, Taipei, 2006.

Urry, John. *The Tourist Gaze: Leisure and Travel in Contemporary Societies*. London: Sage, 1990.

Wang, Po-Chun. "The Imagination of Clusters: Implementation and Policy Making of HuaShan Creative Park." Master's thesis, Taipei Art University, 2011.

Wen-ting, Tsai. "A Green Legacy - The Japanese Houses of Taipei." *Taiwan Panorama*, May 2005, 70.

Yan-chih, Mo. "Taipei to preserve historical Japanese-era buildings," *Taipei Times*. 22 January 2009.

Yeo, Wei Wei. "City as Theatre: Singapore, State of Distraction." In *Postcolonial Urbanism: Southeast Asian Cities and Global Processes*, edited by Ryan Bishop, John Phillips, and Wei Wei Yeo, 258-61. New York and London: Routledge, 2003.

Yeoh, Brenda, and Li Kong. "The Notion of Place in the Construction of History, Nostalgia and Heritage in Singapore." *Singapore Journal of Tropical Geography* 17, no. 1 (June 1997): 52-65.

Yi, Ho. "Preserving the Past." *Taipei Times*. 31 July 2013.

Zheng, Jane. "'Creative Industry Clusters' and the 'Entrepreneurial City' of Shanghai." *Urban Studies* 48, no. 16 (December 2012): 3561-82.

Zukin, Sharon. *The Cultures of Cities*. Cambridge, MA: Blackwell, 1995.

— "Gentrification: Culture and Capital in the Urban Core." *Annual Review of Sociology* 13 (August 1987): 129-47.

— *Loft Living: Culture and Capital in Urban Change*. London: Radius, 1988.

CHAPTER FIVE:

The Kitschy, the *Shanzhai*, and the Ugly: Creating Architectural Utopia in Contemporary Chinese Cities
Lu Pan

Introduction: *Xiaolingtong's Travels* and the Contemporary Chinese Spatial Utopia

In Ye Yonglie's *Xiaolingtong's Travels in the Future* (1980), the first science fiction novel published after the Chinese Cultural Revolution, the protagonist, a young journalist named Xiaolingtong, finds himself mysteriously landing on a ship that is sailing to the "City of the Future" (Figure 1).

Figure 1: Cover of the comic strips of *Xiaolingtong's Travels in the Future (1980)*

A large, multi-generation family warmly welcomes him and guides him around the city while introducing him to new technological marvels, a modern urban environment, and a new lifestyle unknown to Xiaolingtong's time, which is supposedly late 1970s China. The "City of the Future" is undoubtedly a utopia, where technology has a major role in material accumulation, energy provision, labor, entertainment, education, and extension of human lives. Xiaolingtong finally rides back on a rocket and returns to where he embarked on the ship to the "City of the Future." In the story, he is an outsider, or rather a contemplator of his new experiences during his trip.

The story and its subsequent comic version resonated among young Chinese readers throughout the 1980s and beyond. Estimates indicate that more than sixteen million first edition copies of the book were sold. The book succeeded not only because it satisfied the craving for a depoliticized fantasy world in post-revolutionary China, but also because it rekindled faith in a future utopia that was promised by the Revolution. After all, the projects of Communism, in many ways, are themselves products of a sci-fi-styled imagination of a future utopia. Although Ye's *Xiaolingtong's Travels in the Future* was first published in 1978, it was actually written in 1961. This detail reflects the similarity between the literary genre and the Communist ideological project. It may not be coincidental that the book was written right after the end of the Great Leap Forward (1958 to 1960), during which propaganda posters commonly displayed visual presentations of giant fruits and vegetables, just like those in the "City of the Future." While the reason for the late publication of the book is unknown, it seems that even after the reform began, the Communist or modernist vision of utopia made possible by science and technology has remained. This vision persisted despite the disastrous consequences of the Great Leap Forward and the shift of the nation's development during the Cultural Revolution. It continued to live on probably because much of its promises have not yet been realized.

I therefore propose that Ye's *Xiaolingtong's Travels in the Future* is a symbolically renewed starting point of Chinese spatial creativity after the Cultural Revolution. This starting point provides a historical explanation of a possibly different popular imagination and memory of the concept of "creativity" in the contemporary Chinese context. While 1980s China might have retained this creativity on a political or spiritual basis to achieve the utopian world, it was not until the failure of the student movements in 1989, which put an end to the prevailing revolutionary romanticism, that a new series of creative reconstructions of the material world finally took place. The change is an internal one or a Jungian "metanoia" that tries to self-repair, heal, and consequently be unaware of existing wounds. The metanoia aims to accelerate time to compensate for what has been lost; in the case of China, what was lost was time for national modernization. The idea of conquering time remains, first of all, consistent to the Communist idea of historical time. Second, time can be conquered by producing new objects and concepts on a large scale. Again, Ye Yonglie's story provides an alternative way to produce the new: instead of traveling in time, Xiaolingtong travels in space. Therefore, it is not surprising to see that, since the late 1980s, China has been undergoing tremendous transformation in its spatial order and architectural styles. "Chai," or demolition, is a top keyword that describes Chinese social developments. Construction and reconstruction take place on a massive scale and involve the creation of large architectural artifacts. Thus, a utopia that is characterized by extensive innovations is being shaped spatially and temporally to make up for the time lost. This creation of a utopia is propelled by the new socialist market economy system. As in the city that Xiaolingtong encountered during his travels, today's China *is* a "City of the Future."

In most discussions about the development of China, the question is whether China's architectural creativity, if such a thing exists, is largely due to China's desire to demonstrate to the world its rapid economic growth or to the rise of the Chinese middle class, of Chinese cultural appropriation of "the other" or "the alien," of consumer behavior, and of controversies over intellectual property. Within China, *Shanzhai* (a Chinese word that means roughly copycat or knock-off) architecture and townscapes, many of which are imitations of Western architecture and monuments, are denounced by the party newspaper as "a sign of a lack of cultural confidence" or a kind of "blind worship of foreign goods and ideas."[1] Outside the country, a typical Western interpretation would regard the current architectural duplications and innovations as "monumental assertions of China's global primacy" (Carlson, 2012) or a way of "domesticizing foreign culture" (Marinelli, 2013). Some link the copying of exotic landscapes to China's ambition of global primacy; this copying is, therefore, said to exhibit a kind of triumphalism. Bianca

Bosker, the author of *Original Copies: Architectural Mimicry in Contemporary China*, believes the current urban construction in China is a rejection of the architectural practices, assuming they existed, of the Mao era, as these practices are deemed to be symbols of being "poor, uncomfortable, regressive" (Bosker, 2013: 80).

One of the major aims of this paper is to revisit the most prevalent observations of Chinese architectural practices, which are seen as outcomes of the tension between China and the West, or between China's socialist past and its post-socialist present. I argue that these architectural trends in China are not necessarily a display of its power and ability to conquer the world, nor are they a break from the country's Communist past. In my view, architectural designs in China reflect an upgraded Communist utopia that is not interrupted but instead is improved by, for example, continuity of Communist sharing in visual terms, further acceleration of historical time (space travel as time travel), and the implementation of capitalism as part of the Communist project. New buildings use space rather than time to create a perpetual zero hour of history, not through war or natural disasters but through (re)construction. This perpetual zero hour puts the Communist utopian project back on track.[2]

In relation to this view, I also aim to redefine the idea of creativity and "the new" in architectural practices in post-1989 China. I will examine these practices through the lens of what I call post-Communist "visual utopias." I define "visual utopia" as a spatial configuration that only provides a spectacle that fulfills the ideals of a utopia with signs, symbols, and icons. In the case of contemporary China, this utopia is a substitute for the completion of a Communist utopian project that can be characterized by egalitarianism, shared property, and progressive historical time.

Correspondingly, I will discuss three cases of such visual utopias in China's current socialist market economy, which paradoxically accommodates two inconsistent ideologies in one reality. First, I will start with the decades-long trend of building miniature theme parks all over China, which began in the late 1980s. With kitschy reproductions of iconic landmarks both in and outside of China, these miniature parks provide visitors an egalitarian experience. Second, I will focus on how the logic of *Shanzhai* architecture can be understood as a means of creating a "utopia of constant global mobility" (Groys, 2008: 108). In the era of global tourism, "authentic" copies of towns and architecture in the West or from past eras in China aim to create an aura and a sense of sharing through spatial reproduction. Finally, when more and more "hyper-modern" urban artifacts with futuristic forms are built in China, the Chinese public, with the help of the Internet, uses self-organized public polls to reject these "extraordinary" buildings that seem to be ahead of their time. I argue that these cases may imply more underlying desires that are suspended or repressed in the complicated political and economic situation in post-1989 China. At first, when the Chinese urban elite designed and built visual utopias, these architectural works seemed to provide a remedy for the unrealized promise of two utopias for the Chinese public: a Communist utopia of equality and a capitalist utopia of market freedom. Later, however, these top-down ideals of urban forms are no longer shared by the Chinese public, which suggests significant changes in the politics of urban space in contemporary China.

The Kitschy World of Miniatures

The craze for miniature parks is not a Chinese invention. The first miniature parks were established in Europe when people were nostalgic for the industrial age and modern tourists sought a panoramic experience. Bekonscot Model Village in England was created in 1929 to showcase scenes of rural England; today, it depicts rural England as it was in the 1930s. Madurodam, a Dutch version of Bekonscot, opened in 1952 and features the canal houses of Amsterdam, the Alkmaar cheese market, and parts of the Delta works.[3] In China, widespread construction of miniature parks began in the late 1980s, and miniature parks eventually became highly popular tourist attractions. Splendid China (*jinxiuzhonghua*), the first miniature park that features famous Chinese landscapes, opened in Shenzhen on November 22, 1989, and drew more than three million tourists during its first year. Opened in 1994, the Window of the World shows famous Shenzhen landmarks and scenes from all over the world. It recovered its investment in less than three years. The total profits of these theme parks were estimated to be over three billion RMB. Chongqing opened its own miniature park in 1992, and Beijing's World Park was launched a year later. Other cities, such as Chengdu (1994), Xi'an (1995), and Changsha (1997), followed suit.

These parks capture the essence (*jinghua*) of Chinese and world civilizations, thereby enabling tourists to both enjoy such civilizations from a distance andt experience them intimately. Reproductions of landmarks in the miniature parks can be described as kitschy. Kitsch, although a widely debated term, is generally defined as "in bad taste" and refers to the tasteless imitation of a thing that usually evokes in its viewers a universal sentiment of beauty. In miniature parks, kitschy reproductions evoke memories of the past and juxtapose spatial existences: the Little Mermaid from Denmark sits near Venice's San Marco Plaza, with the Sydney Bridge at the back. This surrealist co-habitation is largely decontextualized and is a mere cluster of stereotypical cultural icons.

Although largely discussed in a capitalistic context of commodity fetish, kitsch can also be found in Communist conditions.[4] Temporally, capitalist kitsch is nostalgic and born out of fear of constant change that occurs with modernity. It aims to overcome anxieties over change and alienation. In modern society, constant introduction of new things increases people's tendency to forget traditions and one's past. Kitsch can be likened to childhood memories, as it allows people to continue to experience the past (Benjamin, 2007: 183). Kitsch may provide subtle clues to the past that have been lost in a present full of changes and a "poverty of experience" (Benjamin, 2005: 732).[5] In contrast, Communist kitsch is forward looking. With the upcoming arrival of a utopian world, Communist kitsch attempts to eliminate any uncertainty or hesitation toward the present. Like capitalist kitsch, Communist kitsch also aims to ensure linearity and endurance of historical time. Both promise accessibility and universality, which indicates a utopian sentiment.

In Chinese miniature parks, kitsch objects are embedded in both temporal directions, as can be seen in the success of both Splendid China (and also its Folk Culture Village) and Window of the World. For one thing, China in the early 1990s urgently needed to revive the national imagination after a long period of isolation during the Mao era, the economic reopening under Deng, and the unsuccessful pro-democratic student movements. Splendid China was established at just the right time. The miniature representations of the beauty of "Splendid China" imbue a sense of national unity based on land, ethnicity, folk culture, and cultural heritage rather than on ideological uniformity or class-consciousness. For another thing, at a time when foreign travel was inaccessible to most Chinese in the 1990s, such miniature parks enabled them to temporarily fulfill the dream of traveling abroad without the restriction of visa controls and without having to spend a large amount of money (Figure 2).[6]

Figure 2: Miniature Great Wall in "Splendid China" in Shenzhen

Thus, the miniature parks conjured an egalitarian visual utopia for the Chinese public in the 1990s. Despite Chinese revolutions, an egalitarian world without class, with social equality, and with freedom of mobility has not yet been created. However, the parks somehow provide an equal opportunity for the Chinese public because they allow people to enjoy and appreciate a smaller version of a visually egalitarian world. This visual utopia is realized by spatial coevalness, which means in the first place that visitors aim to experience both immobility and mobility. As the representational space of immobility, the miniature parks illustrate the paradox of post-1989 China. On one hand, images of the outside world enter the everyday life of the Chinese and increasingly circulate in mass media, theme park, commodity packaging,

Politics and Aesthetics of Creativity

and domestic/public decorations. They portray a utopia without travel restrictions. On the other hand, the kitschy reproductions reiterate a sense of immobility for the majority of Chinese citizens, as shown in Jia Zhangke's film *The World* (2004), which is about a group of migrant workers in Beijing's miniature world theme park. Miniature worlds fulfill the utopian dream of mobility. The promise of opening up and "being in line with" (*jiegui*) the world in the post-revolutionary years was first played out in these parks before it was realized. Traveling around China or the world is highly appealing to people who feel bored in the meantime. Illusionary mobility normalizes stagnancy as an enjoyable experience.

Second, this spatial coevalness is realized by the relationships between the real and the kitschy, or by a kitschification of the reality. Taking photographs is essential in visits to miniature parks as well as in "Chinese style" tourism. Interestingly, as in the beginning of Michelangelo Antonioni's documentary *China* (1972), Chinese people lined up for photographs in front of Tiananmen Square as part of a revolutionary pilgrimage or package tour. In post-revolutionary China, people remain enthusiastic about taking photographs before an iconic landmark. Like collecting postcards, taking pictures in front of these artifacts completes the ritual of kitschifying the space, narcissistically "with an atmosphere saturated with 'beauty,' that kind of beauty one would wish to see one's daily life surrounded with" (Călinescu, 1987: 248). Some images from Thomas Sauvin's photo collection project "Silvermine Project" are used as an example to illustrate this kitschification of the reality in Chinese miniature parks. In Sauvin's collection, half a million discarded negatives dating from the 1980s to the early 2000s that were previously owned by ordinary Chinese people were recycled, developed, and digitized. The collection presents a visual world of the photographic norms and fantasies of the Chinese people during a rapidly changing time. At the same time, it may reveal a lot about the way the Chinese people relate themselves with their spatial imaginations. This collection features a series of photos entitled "Around the World in Eighty Minutes," in which an elderly couple visited a miniature park and posed happily at different spots. In one of the images, the Sydney Opera House was on the right side while, surrealistically, two high-rise buildings outside the park were at the center of the photo (Figure 3).

The Kitschy, the Shanzhai, and the Ugly

Figure 3: Around the World in Eighty Minutes: the Sydney Opera House
and Shenzhen's new highrises; courtesy of Thomas Sauvin

In another photo, tall buildings were under construction behind the Thai Grand Palace (Figure 4).

Figure 4: The Grand Palace and Buildings under Construction; courtesy
of Thomas Sauvin

The miniature world is no longer made surreal by the presence of Chinese tourists; it becomes a part of the real world. In this way, tourists unify the outside world harmoniously with the surreal landscapes. Rather than provide an escape from reality, as Disneyland does, the miniature parks reflect the real world and its rapid transformation. As Shenzhen's rapid modernization since has shown, half-finished modern buildings have also become kitsch objects like the miniature world's fake heritage structures. The inclusion of the outside world in the photographs suggests that the border between the banal and the kitschy did not exist. Instead, they are mutually dependent for their existence.

Thus, miniature parks provide the fantasy of a world that embraces empty, homogenous time, which serves as the fresh starting point of social progress. After all, China in the early 1990s greatly needed an image of its present that can explain its break from its Communist past and the dual-ideological present, which is a traumatic moment that might be healed by a borrowed memory. It is not a coincidence that Shenzhen was the first city where the passion for miniature theme parks was kindled. Almost a tabula rasa, the southern Chinese city, which borders Hong Kong, grew from a nameless fishing village to the most rapidly modernizing Chinese city under Deng Xiaoping's opening-up policy since 1978. However, despite its new skyscrapers, Shenzhen had no previous memory that it can share. The city's rapidly modernized cityscape has no references to its own past. The accumulated time in spatial concentration unfolds before the tourists to allow them to reconstruct their lost memories to ensure that the temporal disjunction can be represented and confronted in space. Tracing the history of China and the world could ensure that Shenzhen would also have a place in the historical present. The parks thus provide a "utopian moment *within* the disjunction itself" (Katz, 1998: 12). Here, the mass consumption of kitsch helps to construct the history of ahistory.

Creating Aura with *Shanzhai*

If theme parks in China serve only as a prelude to the construction mania of the simulacra that still lies outside of the real world, the unity between reality and kitsch is extended to the practices of building *Shanzhai* architecture in China's everyday space. Not coincidentally, Shenzhen, the origin and center of China's miniature theme parks, has also become the origin of *Shanzhai* culture. With the rise of *Shanzhai* (literally, mountain-fortress) culture, which is a term that is loosely defined as a behavior of copying or pirating famous brands with its own style, *Shanzhai* architecture is sweeping all over China. Apart from copies of one single building such as Le Corbusier's Chapel Ronchamp in Zhengzhou, Henan Province, or, most recently, Zaha Hadid's Wangjing Soho in Beijing, which is under construction in Chongqing and will be finished more quickly than its Beijing original, more ambitious projects are on the rise; such projects involve making replicas of a whole townscape in the newly developed suburban areas of big Chinese cities. The entire town of Hallstatt, Austria, was duplicated in Huizhou, Guangdong Province (Figure 5).

Figure 5: China's Hallstatt in Guangdong Province

In the suburbs of Shanghai, residential quarters that mimic real or imagined architectural styles in Europe promise to provide an "authentic" taste of small towns in England, Holland, France, and Sweden. A "bigger and better" Manhattan area is being constructed in Tianjin's Binhai District, most interestingly with investment from the Rockefeller Group.

The fact that China's *Shanzhai* practices have garnered so many controversies can be attributed to various reasons, among which are China's open disrespect for "creative" authorship, its somehow exceptional position in the capitalist world, and anxieties due to the potential rise of the country's global power in the West. Indeed, both internal and external conditions of Chinese communism have largely changed. On one hand, China is now described usually as a "post-Communist" society that, to a certain extent, betrayed its Communist doctrines by embracing the capitalist free market. On the other hand, globalization is dragging, more or less, the world into an era of new spatio-temporal configurations. It seems that China, among all the other latecomers to global capitalism, is most eager to prove its new identity vis-à-vis its own past and its major competitor, the West.

In my view, however, the creation of these simulacra spaces does not necessarily describe a dichotomy between China and the West. In the first place, unlike a closed and anti-tourist utopia of high Communism, today's China has also become one of the global utopias of "constant global mobility" that can be found all over the world (Groys, 2008: 108). By this term, Boris Groys refers to a noticeable change in the spatio-temporal configuration of today's city in the age of global tourism. In the essay "The City in the Age of Touristic Reproduction," Groys traces the origin of the earlier utopian vision of city-making as "not only utopian" but a dissociation "from space as it moves through time" (101). In what Groys calls "romantic tourism," he sees sentimentalism over the loss of local culture or "the authentic" in this utopian vision of the city. Travelers who visit a foreign place briefly tend to monumentalize and re-monumentalize it into their gaze. Today's globalization, which blurs the boundaries of the nation state as well as cities at an unprecedented speed and scale, can be regarded as the opposite of the earlier utopian city (102). Tourism in globalization enters its "post-romantic" period: "Rather than the individual romantic tourist, it is instead all manner of people, things, signs and images drawn from all kinds of local cultures that are now leaving their places of origin and undertaking journeys around the world" (105).[7] Cities are copying each other, but an archetypal city that other cities would completely take over no longer exists. By now, "the utopia of an eternal universal order has been replaced by the utopia of constant global mobility" (108). In an era when the subject and the object of the tourist gaze no longer have rigid boundaries, as Groys argued, today's architecture "has now begun to move faster than its viewers" and "since we have all become tourists capable only of observing other tourists, what especially impresses us about all things, customs, and practice is their capacity for reproduction, dissemination, self-preservation, and survival under the most diverse local conditions" (107).

If we then consider the construction of *Shanzhai* architecture as part of this global shift of urban utopia construction, it can also be understood as a dramatic acceleration and concentration of the reproductive practices of these traveling spaces in Groys' descriptions. The reasons for this acceleration and concentration should be further discussed in the context of post-Communist China. On the national level of a post-Communist and globalized China, "profanation" (whose religious implication is totally unfamiliar in the Chinese context) of the original architecture can also be understood as a reaction to the inequality of globalization. While China manufactures most of the goods that are sold around the world, few of them can be consumed in China (Abbas, 2005: 18). At its core, *Shanzhai* connotes the idea of sharing and public ownership that is open to all, without restrictions on temporal and spatial boundaries. As the inequality of globalization has caused a sense of exclusion or a prolonged waiting for recognition, uncertainty—not confidence or triumphalism—has propelled the indulgence in *Shanzhai* architecture. These conditions function as remedies for, or at least appear as, "symptoms" of the rupture between the façade of a free market/body and the still-existing global market hierarchy (19). *Shanzhai* architecture is also not a Disneyland simulacra, which connotes an effort to copy both text and context to create a state of likeness beyond recognition. These are real buildings for use and sale, and are not seen as inferior copies but equal appropriations for their own context. According to Walter Benjamin's idea of "aura," "Even the most perfect reproduction of a work of art is lacking in one element: its presence in time and space, its unique existence at the place where it happens to be" (1968: 220). Based on this idea, an artwork may have aura only when the people who have copies of its images in their hands would travel to see it in a specific space. *Shanzhai* architecture does not aim to gain aura for original architecture, but to lend aura to Chinese space as this architecture and space "come to visit" China. Thus, utopian endless traveling cities and bodies overcome China's alleged lack of authenticity and creativity in today's globalization.

On an individual level, we see in the motivations behind these large-scale *Shanzhai* architectural projects the need to travel through space as a strategy to overcome the temporal delay or stagnancy of the Chinese people's process of class ascension. During the rapid economic boom in the past twenty years, the revival of class society has resulted in significant mobility in class, class-consciousness, and self-identification. This scenario is also accompanied by anxieties (if not crises) about people's current and would-be identities. Within China, different players are advancing the replication mania: local government officials, who are among the most important leaders of these impressive projects during the tenure of their office; real estate developers, who try to sell an idiosyncratic residential environment to their clients; and either a rising middle class or grassroots class of the Chinese public that aspires to climb the social ladder. If theme parks provide the Chinese public with a space to fantasize mobility in immobility, real travel is now possible for the Chinese middle class who are the first and foremost target market of parks with quasi-European buildings or bureaucrats who work in White House-like office buildings. Paradoxically, those who are attracted by these *Shanzhai* buildings are exactly those who are on their upward track from a lower to an upper class. Many of them are not cosmopolitan enough to resist the seduction of the European kitsch, which is used here as a capability to dream and create a utopian vision. Confirming the symbolic connection between European aristocratic interior/exterior and social hierarchy is one of the essential proofs of their upward social mobility. They are creating what Pierre Bourdieu would call taste as "social fractions," a way the cultural capital serves as a mechanism to make distinctions for certain classes in a society. For Bourdieu, such learned tastes and aesthetic preferences may reversely consolidate the social immobility of a class. (Bourdieu 1984) In China, the deliberate display of the "different" symbolic label of one's class similarly reconfirms that he or she is still in the making or at least unsure about it. *Shanzhai* architecture again creates an "aura" for those who are still waiting to have a share in the class to which he or she desires to belong. Therefore, *Shanzhai* architecture also serves as a visual utopia for unfulfilled dreams of sharing, ascending, and owning.

The Aesthetics/Politics of the Ugly

While *Shanzhai* architecture has become the central point of the controversies that concern Chinese architectural creativity, asserting that architectural practices in China are languishing in terms of the common understanding of creativity is totally wrong. If both miniature theme parks and *Shanzhai* spaces in China were constructed on the basis of the already existing "beautiful" urban artifacts to embellish Chinese cities as they undergo a transformation, numerous efforts have also been made to create some of the most impressively innovative buildings in the world. Highly imaginative architecture is sprouting on Chinese soil. However, the Chinese public does not necessarily see such creations as beautiful or aesthetical. Some of the most "creative" buildings are nominated for the "10 Ugliest Buildings in China" poll that has been organized by www.archcy.com since 2010 (jianzhu changyanwang).[8] Images of these new landmarks and buildings are widely circulated and mocked in online communities in China. Online voters soared from 7,522 in 2010 to 28,630 the following year. In 2012, 36,321 participated in the voting. The contest is gaining increasing public attention, which is actually not at odds with the prevalence of black humor in the Chinese Internet community.

The contest is well organized. Nomination, assessment, and selection of these buildings are open to the public, which mainly consists of Internet users and architecture professionals. Details of the contest schedule are posted in a specific subpage. Anyone who provides an image of a "qualified" ugly building will be rewarded with a present from the organizer. The website claims that the rationale behind the contest is to promote the healthy development of Chinese architectural culture. Rather than attacking a particular person or a particular building, the contest aims to curb the frequent emergence of ugly architecture. According to the website, "The 'Contest of the Ugliest' will not stop until our architectural design has reached a truly rational and healthy stage of development. This is the ultimate goal of the ugly contest."[9] A reasonably ugly building should include the following traits:

1. Has unreasonable functions
2. Is unharmonious in relation with surroundings and natural environment
3. Is a copy or *Shanzhai*

4. Is blindly xenophile or modeled after an antique
5. Boosts eclecticism and patchwork of style
6. Blindly imitates
7. Overdramatizes symbolism or metaphorical forms
8. Is bizarre or vulgar
9. Feasibility is forced[10]

I conducted a quick statistical overview of the contest by focusing on the Top 10 lists for each year. The results indicate that buildings are selected regardless of their fame, style, location, and political significance. Among the thirty selected buildings, most are commercial buildings (twelve), seven are public institutions (i.e., university library, museum, and Buddhist temple), four are landmarks such as the entrance of a village, two are residential buildings, and one is an illegal construction. This illegal construction, or *mianzi lou*, is an unfinished building with a fake façade that was constructed by the local government to save *mianzi* (face). Three buildings shared the "honor" of appearing twice on the three lists: the China Central Television (CCTV) headquarters, which is well known by its nickname *dakuchaer*, or "the big boxer shorts," in Beijing (Figure 6); Shanghai Minhang District Court, which is almost a copy of the Capitol in the United States; and the Olympic Sports Center, which looks like a table tennis paddle, in Huainan, Anhui Province. These three buildings properly represent the three most controversial architectural styles in today's China: the international style with a futuristic shape, the *Shanzhai* style, and the uncreatively literal reproduction of symbolism. Chinese Internet users usually use the word *lei* (which means "being shocked as if struck by lightning") to describe the sensory impact of these buildings.

The Kitschy, the Shanzhai, and the Ugly

Figure 6: CCTV Tower or "the Big Boxer Shorts" (Dakuchaer)

Two architectural categories may provide particularly useful hints to the politics of ugliness in contemporary China. Noticeably, the *Shanzhai* style is frequently selected in this contest. Therefore, diverse public opinions toward *Shanzhai* in China need further scrutiny on the function, quality, and ownership in individual cases. The severe distaste towards the use of *Shanzhai* for Minhang Court, for example, has much to do with the very obvious connection between a Chinese national authority and an American one. Criticisms focus on its various problems: blatant plagiarism, sheer vulgarism of the corrupt government officials, and an ugly dome. Some netizens expressed their distrust in a Chinese legal institution that was built in a foreign form. *Shanzhai* architecture is also not considered as a tourist attraction. Huaxi Village's *Shanzhai* U.S. Capitol, which is a less extravagant version of its Shanghai relative, ranked fourth in the 2011 Top 10 list. Huaxi Village of Jiangsu Province is widely known in China as a model "socialist new village" of rapid economic development after the Reform. Its "capitol" does not intend to hide its fakeness. Instead, *mei guo guo hui da sha*, the six Chinese characters for "The U.S. Capitol Building," are displayed on the façade of the building. However, as I mentioned earlier, *Shanzhai* practice is not always geared toward copying the West. The Huaxi *Shanzhai* Capitol, together with the *Shanzhai* Great Wall and *Shanzhai* Tiananmen Front Gate, again attest to the need to contract space into a "model space" that is dedicated to celebrating the glory of the socialist market economy as in Shenzhen, but only in a much more vulgar and *Shanzhai* manner. "Overcreative" new ideas for architecture have also encountered setbacks. Rem Koolhaas's global reputation did not help much in impressing the Chinese public with advanced concepts of space. His design of the headquarters of CCTV, which is the most powerful state media company in China, ranked number one and number three in 2010 and 2011, respectively, on the Top 10 list of ugliest buildings. The *dakuchaer* and all other erotic associations this building has evoked among voters have been long-time laughingstocks in China. Back in 2009, a fire that burned the side building of *dakuchaer* has always been believed to have been started to burn the main building; the incident actually overjoyed the public.[11] The China Pavilion in the 2010 Shanghai World Expo, which was a highly valued global event for the Chinese government after the Beijing Olympics, won the 2011 contest with 6,371 votes. It infuriated the voters both aesthetically and symbolically: people left comments that said this is a Chinese hearth, tomb, or structure that symbolizes the oppressive nature of government bureaus in China or is even an embarrassment to mankind.[12]

Of course, we can easily understand that this contest is no longer about architecture. Just like the dissatisfaction with the Chinese national soccer team or the Spring Festival Gala Show, these grumblings about ugliness are another relatively safe outlet for the people to express their discontent with Chinese social realities. These visual invaders that are so common in Chinese cities plainly speak to a highly visible unevenness in the power relations of aesthetics and, therefore, of politics in China today. This dissensus is a result of various factors: the control of a very small number of urban (political) elites over the visual environment, while the public is largely excluded from the consultation and decision-making processes; an overwhelmingly repressive message being communicated by the widespread existence of these buildings in the cities, constantly reminding the public of its insignificance under such physical symbols of power; the hegemony over the definition of "beauty" by a predominant class that is unqualified to direct public taste; and the use of an outdated paradigm that imposed uniform aesthetics on the public during the revolutionary years. The problems of the social condition and benefit relations have turned into a debate over beauty and ugliness.

Therefore, a utopia of ever-updating visual newness that is defined only by government officials, real estate developers, or a very few architectural experts is being challenged by the public. If making differences is seen as the shortest way to show creativity and garner attention, "ugly" architecture is trapped by a series of paradoxes that achieve ironically reversed consequences for certain people. First, implementing more changes causes these changes to become less acceptable. As Ackbar Abbas argued, "What is disorienting in these and other Asian cities is not so much the unfamiliarity of new places, but the way the coordinates of the old places seem to have shifted, the unfamiliarity not of the new, but of the old" (2005: 246). Visual development is thus not necessarily the most significant crystallization of urban structural change. On one hand, it may only reveal whims of authoritarian wills of power, and on the other hand, "major changes in urbanism are initially registered and grasped at the affective level, in terms of a subjective, experiential response to space" (248). Abbas used Shanghai's newly developed Pudong district as an example: "Pudong, for example, is the district in Shanghai where the majority of new architecture is found, but Pudong is not Shanghai….Paradoxically, it is the visual that can make Asian cities invisible" (248).

Second, as buildings begin to look more extraordinary, people increasingly prefer to see the ordinary, which is perceived as even better than the extraordinary. The desire to produce representationally extraordinary urban artifacts results in a negative impressiveness that is anti-representation. Modern, creative, and utopian architecture that has arisen from the ruins of the old caused ordinary and outdated urban textures to become ruins themselves, upon which beauty is redefined as ugliness. The ordinary or the "common unhappiness" in Freudian hysteria studies then becomes a critical tool to the extraordinary and the different. The unwelcoming attitude toward the architecture despite the political importance or the background of the designer may echo what Abbas described as "a politics of disappointment" rather than a politics of hope. A necessary action is to "recognize (that) objects, states of living, and even the unanticipated consequences of our actions are never quite where we think they should be, and may not end up necessarily fitting our vision of what the world should be or the time frame in which we think it should happen."[13] The utopian image of a city of creativity that turns into a dystopian disaster may encourage us to rethink the relationship among the creative, the new, and the different.

Conclusion: Post-Communist China and the New

I will conclude my observations on architectural practices in contemporary China with the following theoretical questions. How do we assess creativity in China? Are originality and authenticity the only roots of creativity? Can we simply say that Chinese architectural kitsch, copying, and ugliness are not creative? While these questions may be examined through legal, industrial, and cultural frameworks, I am more interested in the larger historical and ideological context that concerns the issue of creativity. Today, the idea of creativity acquires an entitlement of the authentic difference from an exclusive authorship. The narrative of creativity proceeds as an innovative subjectivity invents the "firstness" of an object in temporal order. In the Western context, creativity, which was associated with the artwork of Christianity, is also seen today as a precious, human capability to produce the new. In the discourse of the Marxist criticism of capitalism, for example, creativity is the antithesis of work, which "alienate(s) them from the products of their labor, to reduce them to conveyor-belt functionaries" (Morgan and Ren, 2012: 127). Creativity today also largely suggests the courage and ability to fantasize, imagine, think beyond the current possibility, and make changes and differences. In China and beyond, the creative industry is indiscriminately seen as an updated and productive way to boost the economy.

This view, however, was questioned by Groys against its modernist progressive obsession. In his essay "On the New" (2008), Groys revisited, à la Søren Kierkegaard, the definition of "the new," a concept that is usually equivalent to "the creative." According to Kierkegaard, the real new is not something that we differentiate but is indecipherable and indistinguishable from ordinary objects. In this sense, "being new" is not the same as being different:

> "...a certain difference is recognized as such only because we already have the capability to recognize and identify this difference as difference. So no difference can ever be new—because if it were really new it could not be recognized as difference. To recognize means, always, to remember...but for Kierkegaard the new is a difference without difference, or a difference beyond difference—a difference that we are unable to recognize because it is not related to any pregiven structural code" (2008: 28-9).

While the Kierkegaardian example of the real new is God as "we put the figure of Christ into the context of the divine without recognizing Christ as divine," Groys reminded us that after Duchamp, "the ready-made" which prevails in the contemporary art world may also be seen as the genuinely new because you cannot tell the difference between "the artwork and the ordinary, profane object" (29). In this way, "the production of the new is merely a result of the shifting of the boundaries between collected items and non-collected items, the profane objects outside the collection, which is primarily a physical, material operation..." (34). The existence of museum space that defines exhibited objects as dead also defines living things by excluding them. A museum is a place where the new is collected and exhibited, but paradoxically, the new immediately turns into the ordinary once it enters the museum space. Accordingly, the extraordinary or the different can only be found in the real world.[14]

If the difference can only be said as "the old new" that is not "the real new," creating the new is not about producing authenticity and exclusivity because the real new is hidden, invisible, and only made visible through the change of space. In this vein, all discussions on the question of authenticity, originality, and the real may need to be revisited. The usual belief that authenticity and innovation are temporal issues is challenged and replaced by a spatial one. "The postmodern criticism of the notion of progress or of the utopia of modernity becomes irrelevant when artistic innovation is no longer thought of in terms of temporal linearity, but as the spatial relationship between the museum space and its outside" (Groys, 2008: 34). The interdependence between the original and the copy has created viewers' ownership of its reproduction prior to the birth of the original. In this sense, simulations and reproductions of the original may also change identities and statuses once the original's topological position changes. The claim that no difference between the original and the reproduction exists is, therefore, no longer a question of the dichotomy between the real and the fake, and how the fake can infinitely approximate or finally equate the real. The question is now where you see it. Groys quoted Douglas Crimp's "On the Museum's Ruins" that applies in the contemporary art world:

> "Through reproductive technology, postmodernist art dispenses with the aura. The fiction of the creating subject gives way to the frank confiscation, quotation, excerption, accumulation and repetition of already existing things. Notions of originality, authenticity and presence, essential to the ordered discourse of the museum, are undermined. The new techniques of artistic production dissolve the museum's conceptual frameworks—constructed as they are on the fiction of subjective, individual creativity—bringing them into disarray through their reproductive practice and ultimately leading to the museum's ruin. And rightly so, it might be added, for the museum's conceptual frameworks are illusory: they suggest a representation of the historical, understood as a temporal epiphany of creative subjectivity, in a place where in fact there is nothing more than an incoherent jumble of artifacts..." (2008: 31-2).

Thus, the most interesting thing about the current architectural practices in China is not about distinguishing the self from the other, the original from the copy, or the creative from the plagiaristic, but that the reproductive space is used as forms of creating a kind of post-Communist visual utopia. This utopia is where the immortality of the new in the museum space can be extended to the real world, or the real world becomes a "total museum." One can perform at least three actions to produce the Kierkegaardian new: first, to put the ordinary and the banal into a museum; second, to extend the realm of the museum or shrink the real world; and third, to accelerate time from the different to the new. The three cases I explored here provide the three ways of creating the new. In the first case, the miniature world theme parks render an ahistorical museum space that features spatial juxtaposition of historical signs. These reproductions of the banal and the kitschy images of the real world in parks provide the viewers a utopia in which time is contracted for more efficient spatial travel. In the second case, the reproductions overflow from the quasi-museum space of theme parks to the tactile environment of everyday life. The *Shanzhai* architecture and cityscapes in their life-size scales erode the "original" space of these Western architectural artifacts by making them travel only through reproduction. This scenario creates a utopia of the perpetual mobility of the new in an auratic exhibition space. Finally, if the different or the extraordinary would become the ordinary over time and then become the new in historical time, the change from the different to the new can only be accelerated through spatial reproduction, simulation, and copying. The "creativity" of the different ultimately transforms into reproductive capability. Through reproduction of the different, the different will eventually become the new and the ordinary. The creativity of a visual utopia can then be understood as an accumulation of the old new in preparation for the real new. Thus, visual utopias are created in contemporary China by accumulating the new for the future to surpass time, as previously mentioned. From children's science fiction to the strangest architectural forms, Chinese cities are constantly moving back and forth in time and space to invent the new.

The story of Xiaolingtong's journey ends in an interestingly ambiguous way. In the municipal library, where our protagonist tried to find more information about the past of the "City of the Future," he found that the last few pages of the history book are blank. Instead of foreseeing the future, Ye saw no possibility or necessity to use his wilder imagination. By its invisibility, this utopia is made sacred. The real story of China after Xiaolingtong's journey proves to be a profane illumination of an invisible and maybe unspeakable utopia by various ersatz utopias. Contrary to the common belief, these phenomena are not the interruption but rather the continuation of the Communism project. The visual utopias substitute physical mobility with visual equality; shared right to the space with a reproduction of *Shanzhai* space; and a dramatized progression of historical time with a massive production of visual differences. The paradoxes that are found in the logic and practices in architectural creation, like many other things in China that endeavor to escape from time and space constraints, are explained with the sophistry of "the actual conditions of a country" (*guoqing*). It provides a justification for whatever is "abnormal" in China. The right to abnormality in post-revolutionary China's main official narrative is an excuse to legitimize the postponed realization of the utopian promises. Now, the ersatz utopias that have been largely produced by a joint force of popular imaginations and top-down wills are no longer unanimously seen as desirable, beautiful, or universal. However, if debates over Chinese urban aesthetics aim to create another visual utopia where everyone is expected to have good taste for the appearance of urban environments, we can say that this objective is still a part of the Communist utopia of an enlightened public. This time, however, non-expert political leaders are involved. When we come to the question of how we can fathom the distance and proximity between the new and the different, the ordinary and the extraordinary, or the original and the fake, a utopia of any kind may not provide a sufficient answer.

Notes

1. See People.com.cn. (2013) *Shanzhai Western Architectures Show a Cultural Loss* (shanzhai yangjianzhu, wenhua zaimishi) http://culture.people.com.cn/BIG5/n/2013/0221/c87423-20552320.html, (accessed 27 September 2013)
2. See Groys, 2009: 104: "Passing from the project to its context is a necessity for anyone who seeks to grasp the whole. And because the context of Soviet communism was capitalism, the next step in the realization of communism had to be the transition from communism to capitalism. The project of building communism in a single country is not refuted by this transition, but is instead confirmed and definitively realized. For communism is thus given a historical location not just in space but also in time; that is to say, it becomes a complete historical formation with the possibility of even being reproduced or repeated."
3. See http://www.madurodam.nl/en/ (accessed 16 December 2014).
4. Despite differences in their attitudes towards work and leisure, hedonism and asceticism, class and taste, kitsch is used in both societies as a tool to train the eyes of society's members to adjust to a totalizing visual system—mostly an idea about beauty. Kitsch objects, such as cheap souvenirs and propaganda posters, are repeated as pre-digested clichés in everyday life. To be accepted and favored by the majority, kitsch artists apply a "principle of mediocrity" (Călinescu, 1987: 248). Therefore, kitsch usually speaks to a kind of collective desire beyond individual diversities.
5. For Walter Benjamin, the outdated traditions of the past and the kitsch objects have a deep connection. He uses the idea of experience (*Erfahrung*), an unconscious process that is different from conscious experience (*Erlebnisse*), to reveal the dialectical power hidden in kitsch. See Menninghaus, 2009: 44. See also Benjamin: "Experience is a matter of tradition, in collective existence as well as private life.

It is less the product of facts firmly anchored in memory than of a convergence in memory of accumulated and frequently unconscious data" (1968: 157).

6. Actually, the idea of building miniature parks bears an astonishing resemblance to *Xiaolingtong's Travels*. In Ye's book, Xiaolingtong's new friend, a schoolboy named Little Tiger (Xiaohuzi) from the "City of the Future," tells him about his geography class. The teacher took students on a world tour by atomic jet within a day: they departed at 8:00 a.m. and arrived in Paris at 8:30. After a short stay, they were in London by 10 a.m. They then flew to Egypt and had lunch somewhere near the Pyramids. In the afternoon they flew to the South Pole and ended a day's class. In the evening, they went home for dinner. The miniature parks are perfect recreations of the world travel fantasies described in science fiction.

7. Beijing Silvermine

8. This entails a series of collapses of formerly rigid binaries: tourists/residents, gaze/being gazed at, moving/lodging, departure/arrival, high/low, universal/specific, Western/non-Western, many/few, similarity/dissimilarity, copy/original, urban/countryside are now either equivalent or interchangeable. The mobility of more and more people around the world marked the beginning of a new era of total tourism where, most significantly, "the utopian impulse has shifted direction—acknowledgement is no longer sought in time but in space: globalization has replaced the future as the site of utopia" (Groys, 2008: 106).

9. After the unexpected popularity of the contest among Chinese internet users, the contest for "The Ten Ugliest Street Sculptures in China" was launched in 2012.

10. See Archcy.com, (jianzhu guanchawang) "Selection Criterion for Contest of the Ten Ugliest Buildings," http://www.archcy.com/activity/activity_online/ugly2013/8936a2609c1d8382 (accessed 27 September 2013).

11. Ibid.

12. See blog article, "CCTV's 'Pornographic Building' is the symbol of the total failure of China's Cultural Strategy since the thirty years' open-

up Policy," http://blog.renren.com/share/319103818/13781487189 (accessed 27 September 2013).
13. Archcy.com, "The Chinese Pavilion at Shanghai's World Expo," http://www.archcy.com/votes/2011/new/385 (accessed 27 September 2013).
14. Johannesburg Workshop in Theory and Criticism, "Wrecking Theory or Politics? Ackbar Abbas on Wrecked Objects and Poor Theory," http://jhbwtc.blogspot.hk/2011/07/wrecking-theory-or-politics-aackbar.html (accessed 28 September 2013).
15. See Groys, 2008: 33: "The successful (and deservedly so) mass-cultural image production of our day concerns itself with alien attacks, myths of apocalypse and redemption, heroes endowed with superhuman powers, and so forth. All of this is certainly fascinating and instructive. Once in a while, though, one would like to be able to contemplate and enjoy something normal, something ordinary, something banal as well. In our culture, this wish can be gratified only in the museum. In life, on the other hand, only the extraordinary is presented to us as a possible object of our admiration."

Bibliography

Abbas, Ackbar. "Das Große Fälschen." *Böll. Thema*, Ausgabe 2 (2005): 16-19.

— "Das Große Fälschen. "Fake Globalization." In *Other Cities, Other Worlds: Urban Imaginaries in a Globalizing Age*, edited by Andreas Huyssen, 243-64. Durham, NC: Duke University Press, 2009.

"Supercities and Mega-Migrations: China's Urban Futures Conference Presentation." Columbia University, Wood Auditorium, Avery Hall, NY. 11 November, 2011.

Benjamin, Walter. *Selected Writings*. Eds. M.P. Bullock, M.W. Jennings, H. Eiland, and G. Smith. Vol. 2. Cambridge, MA: Belknap Press, 2005.

— "Traumkitsch." In *Kitsch: Texte und Theorien*, Vol. 18476, edited by U. Dettmar and T. Küpper, 181-3. Stuttgart: Reclam, 2007.

— "The Work of Art in the Age of Mechanical Reproduction." In *Illuminations*, 1st ed., 217-51. New York: Harcourt, Brace & World, 1968.

Bourdieu, Pierre. *Distinction: A Social Critique of the Judgement of Taste*. Cambridge, Mass: Harvard University Press, 1984.

Bosker, Bianca. *Original Copies: Architectural Mimicry in Contemporary China*. Honolulu: University of Hawaii Press, 2013.

Călinescu, Matei. *Five Faces of Modernity: Modernism, Avant-garde, Decadence, Kitsch, Postmodernism*. Durham, NC: Duke University Press, 1987.

Carlson, Jack. "China's Copycat Cities." *Foreign Policy*. 29 November 2012. (accessed 29 November 2012).

Groys, Boris. *Art Power*. Cambridge, MA: MIT Press, 2008.

— *The Communist Postscript*. London: Verso, 2009.

Katz, Marc. "Rendezvous in Berlin: Benjamin and Kierkegaard on the Architecture of Repetition." *The German Quarterly* 71, no. 1 (Winter 1998): 1-13

Marinelli, Maurizio. "Domesticating Foreignness in China: The Transnational Politics of the Copy and the Real." Abstract of paper presented at ICAS 8th International Convention of Asia Scholars, Macau, 24-27 June 2013.

Menninghaus, Winfried. "On the 'Vital Significance' of Kitsch: Walter Benjamin's Politics of 'Bad Taste.'" In *Walter Benjamin and the Architecture of Modernity*, edited by A. E. Benjamin and C. Rice, 39-57. Melbourne: Re.press, 2009.

Morgan, George, and Xuefei Ren. "The Creative Underclass: Culture, Subculture and Urban Renewal." *Journal of Urban Affairs* 34, no. 2 (2012): 127-30.

Ye, Yonglie. *Xiaolingtong's Travels in the Future*. Liaoning: Liaoning Meishu Chubanshe, 1980.

CHAPTER SIX:

(In)Dependence, Industry, and Self-Organization: Narratives of Alternative Art Spaces in Greater China
Elaine W. Ho

When director of the Barcelona Museum of Contemporary Art (MACBA) Bartomeu Marí Ribas visited Beijing in the spring of 2013, he shared an account from his journey to China: during the casual protocol of friendly chit-chat with the passenger next to him on the flight, he was surprised by an unassuming inquiry on the status of the "art industry" in Spain compared to China.[1] This was not because of any particularly great lapse between the two, but because Ribas had never heard the term "art industry," the crux of his ignorance lying specifically in the juxtaposition of visual arts with the modes of industrial production. This naïveté seems like a form of deliberated idealism considering Ribas's position embedded within a high institution responsible for an important part of the circulation of art and cultural production in his own country and internationally. It asks us to recall a pre-Fordist mode of art-making, *pre*-institutional critique and *pre*-producer. But especially when considering the context of this journey to China, a transnational gathering aimed at feeding the engine of emergent Chinese art institutions, we can only claim Ribas's linguistic denial as weak at best. Where he may have attempted a form of resistance to art's full assimilation into the chain of creative industries, Ribas remains ingenuous to the mechanisms that have allowed this excursion to China—an interconnectivity between the realms of government policy, capital and creative labor which, in the Chinese language at least, falls under the extremely common terms 文化艺术企业 *wenhua yishu qiye* "art and cultural enterprise" or 文化艺术产业 *wenhua yishu chanye* "art and culture industry."

In fact, Ribas's visit was part of a celebration of art and multitude in Asia, a conference held in conjunction with the newly announced Multitude Art Prize, co-sponsored by the Multitude Foundation and Wuhan Art Terminus (WH.A.T.). Not incidentally, both organizations are directed by cultural producer Colin Chinnery, who during the conference discussion modestly professed his motivation to seek guidance from the invited experts representing established institutions such as the Mori Art Museum in Tokyo, Museum of Contemporary Art in Antwerp, the Museum of Modern Art in Ljubljana and, of course, Ribas's organization, MACBA. Chinnery's one stone effectively moves private funding in service of an international platform of cultural knowledge production, inadvertently or not also placing China at the center of the action, both as a host for a global intellectual sphere and as a potential barometer of accumulated expertise in the form of a new art institution. The Wuhan Art Terminus, still under development, is one of several new, privately-owned art centers in the second-tier city, including the K11 Art Village and OCAT Wuhan, all supported by the incestuous dynamics of state-owned enterprise and property developer that befit the vocabulary of "art industry." These top-down expansions from the cultural centers of Beijing and Shanghai into the second-tier cities are part of an explicit schema of socio-economic development marked by urbanization and shifted emphasis towards service sectors and knowledge-based economies, including the arts. This is the contextual basis for the spectacular unfolding of the Chinese contemporary art industry in its making, offering the contemporary artist and art worker a wealth of resources and opportunities not so easily secured since the era of revolutionary art organized as labor in service of the party. But it is this correlation—from cultural propaganda to the drawing of socio-capitalist lines of production—from which the map of the mappable has already been drafted. Where the directive for Chinese "soft power" has already been implemented, the need for another tracing of the creative industries, even if adorned with "Chinese characteristics," has already been undone.[2]

New attempts to define and expose uncharted territories have only time and again found themselves in blind service of hegemonic structures. The "natural" character of Chinese characteristics, and even its introduction into the contemporary lexicon, has been carefully handed down as a predetermined strategy and ideological principle based upon economic development. The fantastic rise of the Chinese art market parallel to such top-down molding in the last thirty years may be seen in certain senses as part of the success story, but the increasing complexity of socio-economic relations grounding contemporary artistic creation also appears to follow a path of the mappable, with keywords such as added-value, cultural district development and art superstars highlighting its legend. Underneath this map, however, is the enormous fissure between such top-down initiatives and another topology of grassroots intervention. Marked by ruptures in physical space and an ongoing play between autonomy and heteronomy, the perspective of a so-called "alternative arts practice" in China offers instead an image of the "unmappability" circumscribing a political, cultural, and economic flux. The following analysis seeks, therefore, to embrace the very contradiction that occurs with its production. Case studies of several artist-initiated practices will explore a politics and aesthetics in the making; that is to say, one which has yet to be, or one that defies the certainty of mapping and an art market categorization. What emerges therein is a politics of exception, whereby the complexities of Mainland sociopolitics cannot delineate a formula for artistic production but only be navigated as a realm of singularities, affects, and encounters. If this is a euphemism for Chinese characteristics, it is one that approaches an alter-linguistics similar to Ribas's incomprehension of an art industry. It remains a weak voice within the dominant landscape of artistic entrepreneurialism, and it is one both bound and enabled by such reified structures. But the particular circumstances which have led to the creation of spaces such as Womenjia Youth Autonomy Lab (Wuhan), WooferTen (Hong Kong), Lijiang Studio (Lijiang), and HomeShop (Beijing) exemplify such spontaneous modes of artistic practice. Their genealogies are traced here to highlight a micropolitics embedded within and counter to the prevailing forces of a socialist market economy. It is from within these structures that a meta-analysis will look beyond these groups as a pegged phenomenon and rather as a series of narratives involving practices, engagements, and relations that undo a common thinking of resistance.

Womenjia Youth Autonomy Lab: A Story from Sanctuary to Confinement

Figure 1: In September 2011, Beijing musician Xiao He and Wuhan-based Yao Shisan (*pictured*) held a series of live concerts at Womenjia. (photo courtesy of Womenjia Youth Autonomy Lab)

Since 2008, Womenjia Youth Autonomy Lab has operated as an open house, meeting place, music studio, and library for activists, artists, punks, and travelers, among others. Nestled amidst one of the increasingly rare undeveloped sections of Wuhan, this spatial experiment of leftist and anarchist culture is flanked on one side by farm plots and on the other by the city's botanical garden. This illusion of idyllic sanctuary can be attributed to its geographically peripheral location, which means low rental costs and the uncertain degree of autonomy offered by sociopolitical marginalization. Being less centrally located, and therefore less accessible and less visible, creates less direct provocation for policing authorities; but amidst a regulatory environment of relational repression, pressure is placed on the landlord to kick the so-called "reactionaries" out.[3] Womenjia is thus no sanctuary from the police state, and their activities, including citizen journalism, public screenings, and the hosting of research and dialogue on sensitive topics such as Lesbian, Gay, Bisexual, and Transgender (LGBT), and urban renewal issues, have been continually questioned and monitored by surveilling bodies. This unwieldy relationship to spatiality as both refuge and confinement is exemplified in Womenjia's mixed identity—from one perspective an open door to the likes of young homosexuals who have been ousted from their family homes, and from another, it is exactly the resulting haphazardness of the community that traps the space within, inhibiting a broader dialogue with others. The energy and maintenance required to host a large space with continued traffic from "social refugees" who may share none of the sociopolitical motivations by which Womenjia was founded overwhelms the focus on more specific outputs of artistic and social practice. Thus, initiator Mai Dian's insistence on maintaining free housing in the manner of the European squatter movement from which he was inspired to begin Womenjia is a tremendous obstacle to the work they can do. Amidst the reality that such an occupation would not be tolerated in China in the same way that certain legal frameworks allow it to exist in Europe, Mai has ultimately taken sole responsibility for sustaining the albeit low, but not negligible financial costs. What he claims as trying his "best not to 'benefit' from a so-called 'radical politics'" is a crucial gift that subverts the relationship between economics and space; but it is a practical burden, and at times of internal conflict becomes a question of rights over a stated commons.[4] The common misunderstandings of an anarchist platform in the form of a dilapidated concrete structure housing punks and other outcasts has led, on occasion, to having to expel visitors from the space. And while this does not necessarily contradict from an anarchist practice, it does highlight the ambiguous and difficult margins between sanctuary and confinement. The "public ruins" that Womenjia chooses in form and voice are a commitment to autonomy by way of distancing oneself from the authoritarian-enforced mechanisms of progress as overhaul (Mai, 2014). The primary work, then, is the labor of self-maintenance that is Womenjia itself.

This struggle lies at an awkward position that is neither urban nor rural and perhaps, neither art nor activism. The hesitation to produce, either artistically or via direct political dissent, may be a result of the particular juxtaposition between stunted efficacy under group consensus and Mainland authoritarianism. However, as Taiwanese artist and activist Kao Jun-honn has described it, the project of Womenjia is indeed not about production and producing new relations, but about "establishing from within a complicated structure an activist, subversive and educational zone and space of exchange" (2013).

Lijiang Studio: "Contemporary Art Episodes in Rural China"

Figure 2: The He family's grandmother helps a visitor try on traditional Naxi dress in their home. Behind, the mural painted by artist and project curator Lisa Li illustrates the He family tree. (photo by Elaine W. Ho)

Further remote from Wuhan's East Lake outskirts is Lijiang Studio in Yunnan province, ironically self-described by its founder Jay Brown as a "(voluntary) refugee camp" for itinerant artists.[5] What began thus in 2004 as a residency program for Chinese and foreign artists seeking an alter-space for thinking and learning has evolved into long-term relations between artists and a local Naxi family that helps to host and maintain the space. The studio's location outside the old town in a village of the Naxi, Lijiang's dominant ethnic culture next to Han Chinese, determines the greater part of its relationship to the environment and community. While Brown's long-time residence in China may not fully diminish his position as outsider, his adoption by the He family with which the studio shares space creates a different kind of internalizing process to nurture artistic creation. The Naxi villager as "host" to visitors coming to Lijiang Studio generates an ambivalent relationship of hospitality and servitude (the He family prepares rooms and meals for guests, as well as coordinates local visits), but Brown's curatorial response encourages projects that are formed from the local environment and provide returns in service to the village. There is, admittedly, an ambiguity regarding the nature of the trade within this cultural exchange, but it is exactly the inability to quantify affective relations that offers Lijiang Studio another mode of resilience.

A particularly rich example is the *New Countryside, New Landscape* mural project initiated by artist Lisa Li in 2008. The title references the "new socialist countryside" directive from the central government, aimed at modernizing rural areas diminishing the incredible Gini coefficient figures that mark China's gap between urban wealth and the impoverished countryside.[6] Li's curatorial statement touches upon the ambiguities presented by such a policy, whereby Lijiang tourism and the fading of traditional farming practices create a dilemma for the cultural status of contemporary rural life. Her light-hearted response to this backdrop invites artists from the city "to help re-imagine the role of art in the countryside, which will add a publicly visible element to the various types of nonsense we have added to the village already". (Lijiang Studio, 2013). "Nonsense," here, is Li's clever acknowledgement of the tenuous concerns lying at the base of a foreign-supported platform transplanted into the village, and perhaps another aside to the nonsense of "new-style farmers" and "recycling agriculture" that tagline the new socialist countryside.[7] But opposed to the planned economy of rural reform, the Lijiang Studio mural project engaged Jixiang "Auspicious" Village with the light hearted intimacy of Li's personal efforts to meet and collaborate with local inhabitants for the painting of their homes. Over the course of three years, approximately twenty artists were invited to participate, and results ranged from pleasing portraits of Naxi families in domestic interiors to traditional landscape paintings and works of a more monumental nature, using the entire exterior wall of a building. Some of these included more controversial abstractions that led to occasions of tension between artists and villagers.

Figure 3: A landscape mural painting brings together the collaborative efforts of Manchu artist Na Yingyu, local Naxi artist Mu Wenzhang and Han artist Hu Jiamin. (photo by Elaine W. Ho)

In 2011, one year after the end of the project, Li returned to Lashihai and found that the murals had evolved, not only due to material deterioration, but in their owners' eyes as well. Even those previously most dissatisfied with the project became accepting and were interested in participating in future Lijiang Studio activities, and this progression can be read here as the continuously fluid nature of the discourse on community and social relations. Lijiang Studio's position in Lashihai is an aesthetic intervention in the countryside; but it is one that, much like the role of storytelling in traditional cultures, nurtures over time and binds rather than severs. Whether as memory among the participating groups, a story from the project documentation or in the present analysis, what emerges in the place of a faded mural is not necessarily a transformative sociopolitical relation, nor is it an idealized community. If anything, the re-imagination of art occurs here only in its occurring as a renewed potential. And that is to say, art here is navigated anew each time, as a dialogical practice of subjectivities rather than as a closed package of experience.

WooferTen: Hijacking and Community as Temporalities

Figure 4: WooferTen member Irene Hui takes WooferTen community-made bags and t-shirts out onto the sidewalk to raise funds for the Hong Kong dockworkers on strike in the spring of 2013. (photo by Elaine W. Ho)

If we are outlining each of these spaces in relation to modes of independent arts practice in China, Hong Kong's WooferTen presents the most established of the four case studies in the sense that the form of organization arose, next to the collective efforts of its initiators, in specific alliance with government funding from the Hong Kong Arts Development Council (HKADC). Humorously rumored to be the Arts Council's "most regrettable decision ever," the original group of ten artists, curators, and educators have been explicit about their ambivalent relations to the artistic economy of Hong Kong. From its inception, WooferTen's activities have been artistically punctured by an attitude of "biting the hand that feeds it," whereby an otherwise strong focus upon the Yau Ma Tei community in which WooferTen is based can only rely upon support from a notably more conservative governing body whose chief priorities cannot stray too far from the municipal branding of Hong Kong as "Asia's World City." From distributing stickers with hacked versions of the HKADC logo to hosting the *Siu Sai Gual Bananale* (also known as the *Mini*

West Kowloon Biennale) as critical voice to the dubious development of the over HK$29 billion budgeted West Kowloon Cultural District next to Yau Ma Tei, WooferTen is highly aware of the precariousness of autonomy amidst the flows of global finance. But it is perhaps exactly this dialectical tension that one might say has been enabled by the insistence upon diversity and free artistic expression maintained by the HKADC, where grassroots artistic alternatives such as WooferTen are "tolerated" as a minority expression next to the world-class visions of the M+ Museum under development in West Kowloon.

What is at stake in WooferTen's efforts is a fundamental questioning of local culture and identity in the making of Hong Kong post-1997. With Hong Kong's colonial history making the city a longtime international gateway into China (e.g., in terms of smuggled goods and popular culture), the return to the mainland marks a reversal of such globalization from influx to outflow; Hong Kong becomes a spectacular playground and haven for mainland-managed politico-economic activity (consider the full gamut of implications implied by "Special Administrative Region"). The question of "What is Hong Kong culture?" therefore, is paraphrased in the oft quoted analogy of a frog placed in a pot of water who does not know it is being boiled alive. Yet it is exactly this kind of cynicism that is intertwined with the renewed cultural energy noted by WooferTen's name, literally translated as "living room of revitalization." Cultural critic and curator Kinwah Jaspar Lau notes 2003's climactic July 1 rally, polarized by public opposition to proposed anti-subversion legislation, as the late arrival of "the real 1997" (2012); and the marked increase of initiatives stressing participatory action and local, community-based practice since that time has been continually infused, if not led by contributions from artists.[8]

This background leading up to WooferTen's inception in 2009 is thus a uniquely Hong Kong one, demanded from the specific circumstances that have ignited its artists to consider their roles among a broader motivation for social change. While WooferTen co-founder Chin-wai Luke Ching admits having been influenced by "space hijacking" interventions that he had seen from abroad, one must also consider the particular politics of public space in Hong Kong, a city known as Margaret Thatcher's testing ground for neoliberal policy and where nearly all "public" space—such as pedestrian overpasses, rest areas, and transportation networks—is privately owned (Ching, 2013). The collusion between government and private development results in what Ching calls a "violent policy on space" infiltrating behind the ever-present shopping malls and banners of peace and harmony: "hawkers are driven away in the name of hygiene; the right of abode is seized for the sake of easier ruling; old districts are erased for development" (2008).

The hijacking of space is arguably played in rounds by all actors, from developer to government to citizens, but WooferTen artists' adoption of such confrontational ethics takes on an altogether different temporality since collectively and spatially rooting themselves in an aging, old neighborhood in Kowloon. The politics of hijacking as event are made even more intricate with the slow-kneaded processes of community development and an ongoing exchange of awarenesses and possibilities of art with a non-art public. While humor, "a grassroots neighborhood tone," and an "avant-garde approach" remain, WooferTen has grown to allow for a different nurturing of the processes of "attention, participation, cooperation and interaction in matters of political aesthetic principles" (Lau, 2011). Activities are divided into three categories: a "Special Topics Program" of one-time artist projects; the "Never in Vain Program" of ongoing activities such as a series of workshops entitled "Dare to Teach You if U Ar Willing to Learn" and the monthly Woofer Post community bulletin; and the "Program of Mutual Support," which opens WooferTen's resources to the public as an open platform for collaboration. It is crucial to note, on the other hand, that whatever clarity WooferTen has managed in its organizational representation is a likely adjunct to its means of support. Successful application to a bureaucratic funding body such as the HKADC requires a coherence of form and planning unnecessary for a self-run space such as Womenjia, and it is, ironically, the two-year funding scheme of the Arts Council that has both sustained and strangled WooferTen. Upon the completion of its second term in 2013, the precarious relation snapped, and WooferTen was formally dismissed from the Shanghai Street space in Yau Ma Tei. While the reasons for HKADC's decision are dubious and unclear, supporters point out the politico-ethical contradictions of putting a time limit on the idea of community. Most ironically, then, is WooferTen's great final hijack—itself. Under their self-titled occupation of the space, continued critical dissent and a bureaucratic stalemate, WooferTen has managed to continue its activities on Shanghai Street, albeit struggling under enhanced financial and organizational pressure.[9]

HomeShop: "Temp Space x Time-Plot Ratio"

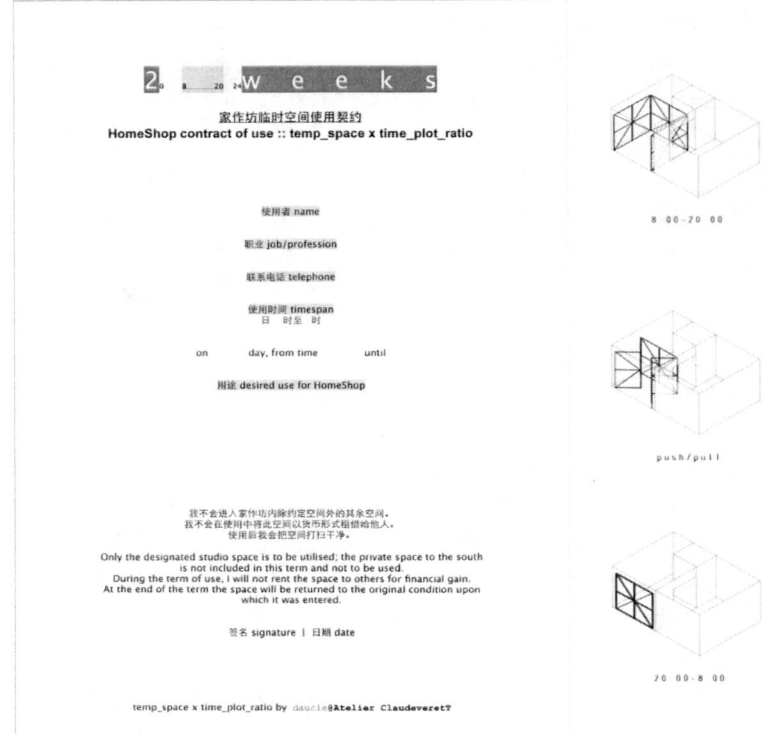

Figure 5: Diagrams of HomeShop and a user's contract made by architect Claude Tao for the "Temp Space x Time-Plot Ratio" project in 2009. (images courtesy of Claude Tao)

Also faced with dissolution in 2013 was Beijing-based HomeShop, the self-described "storefront residence and artist initiative" that had for the prior five years been an experiment with space, community, and the possibilities of artistic collaboration.[10] What began as a personal exploration responding to the array of ambiguities blurring public and private space in China emerged as a collaborative platform defined by these polarities, both in terms of its organization, as an internally binding agent, and representation, as an external one (I will return to this crucial point later). Located in one of the traditional *hutong* alleyways distinguishing Beijing's city center, the daily *mise en scène* here played out in the tensions between old world Beijingers and the influx of young migrants seeking work in the capital, in the residential versus commercial land-use disputes, and in the exponential increase of cars and commercial development in areas that are of historical cultural significance. HomeShop's alternating series of ethnographic fieldwork, artist interventions, and discursive activities were a direct play with the particularities of this village-in-city scenario offered by Beijing's urban dynamic. Its premiere as an alternative countdown to the mega-spectacle of the 2008 Olympics perhaps then set the tone for its future as a marginalized arts practice, and since then, HomeShop balanced a widely diverse community of audiences and participants that distinguished it from the insularity of Beijing's artistic and cultural establishment. This was apparent in its geographic location outside the safe-zone of the city's art enclaves in Dashanzi or Songzhuang, but also in its approach to pursuing forms of representation outside of traditional exhibition practices. Where its glass storefront façade was always intended as a form of display, this was from its inception conceived as an interface for dialogue rather than spectatorship or consumption. The possibilities of the community, therefore, move beyond mere observer or buyer towards neighbor, friend, participant, and collaborator. Earlier activities such as Claude Tao's "Temp Space x Time-Plot Ratio" (2009) and Elaine W. Ho and Fotini Lazaridou-Hatzigoga's "I Love Your Home" (2010) were intentional leaks of the private space of a *hutong* dwelling into the public environs occupied by passersby, onlookers, and potential users. In the former project, the twenty-six square meters of HomeShop was reapportioned as a time slot–distributed free space open for use by anyone from the public. "I Love Your Home" turned HomeShop into a mock real estate agency that freely distributed a catalog of available courtyard spaces, simultaneously offering open access to an alternative map of Beijing real estate and subverting the exploitative practices of property agents.

After two years, the move to HomeShop's second location in a large *hutong* courtyard approximately ten times the size of its previous space served to feed a larger community of possible contributors back into the intimate time-space of living and working together. Financially sustained via the occasional anywhere-capital of grants and a co-working model where participants served simultaneously as supporters, users, and sometimes managers of the space, HomeShop etched out an active but bare survival that finally still remained prey to the devouring mechanisms of the Beijing real estate market. The three-year lease of HomeShop's Beixinqiao location expired at the end of 2013, and in the end, it was a triple rent hike and the western-styled café which took HomeShop's place that had the last laugh.

Of course, this would not have inhibited HomeShop's continued efforts at another location, yet the seven co-organizers in HomeShop's 2013 configuration left Beixinqiao with a mutual agreement to disband the project. The reasons for this are numerous, among them diverging interests and the departure of several co-organizers from Beijing. However, what is more notable in this analysis is the opening up of a broader discussion on the illusions of what curator Carol Lu Yinghua describes as a "self-practice" that attempts "to transcend determination by these boundaries and gradually create self-contained systems based on internal motivation, independent thinking and constant self-appraisal" (Lu, 2011). As part of Lu's ongoing research and discovery of a new mode of "little movements" in contemporary art, HomeShop managed to serve a fleeting moment of interest in the bill for alternative spaces that emerged after 2008 and the subsequent dip in the Chinese art market.[11] As Mai Dian warned, however, "this overlapping of what in China *looks like* different forms of production and political organization" (*italics mine*) actually conceals something hardly different from the territory of immaterial production and hijacked knowledge sharing that characterizes post-Fordist production (2012: 66-7). Both HomeShop's practice (as contribution to a kind of urban cultural gentrification) and its co-optation by institutionalizing theoretical discourses are vulnerable here (Lu has since been named head curator and art director of OCAT Shenzhen; HomeShop reemerged at the end of 2014 with an appearance in a group exhibition held at one of the most established commercial art galleries in the 798 arts district), and it is from this conceptual rut that this paper should begin to necessarily undo itself.

Resistance Undoing or Undoing Resistance

Figure 6: A 2013 participant of WooferTen's annual bicycle procession to commemorate June Fourth. (photo by Elaine W. Ho)

The implication here is that these minor forms of artistic and creative autonomy will be subsumed by their narration into a commodified discourse of resistance. Resistance against whom? The Communist state? Neoliberal capitalism? Art history? Slavoj Žižek is not light-handed in his criticisms of the contemporary Left, precisely because nothing has, of yet, emerged that has not already been renewed and reincorporated by the hegemonies in power. That would in this context most likely include, yes, a totalitarian state, neoliberal capitalism, and even the dogmas of a modernist art historical trajectory that insist upon describable packages of movements and stylistic periods. With the question of "Where are we now?" plaguing cultural studies and the discourse of creative practices, it seems that those protagonists of the new politics of resistance still find themselves, via their critique and counter-cultural strategies, in unwitting service to the dominant disciplinary frameworks. The immaterial labor and knowledge production turns "Where are we now?" into a global network of the avant-garde, propped up by "neo-management" (Mouffe, 2007) and "managerial capitalism" (Žižek, 2012) on one end of the hierarchy and low- or no-wage labor on the other (think of the slew of "glamour-wage" internship positions that populate the creative industries).[12] Both this hierarchy of labor and its byproducts describe a psychology of flailing urgency permeating artistic and cultural production as a whole. Answers to "Where are we now?" seek strategies of resistance only to be continuously fixed as coordinates of creativity for re-appropriation and control. The resulting cat-and-mouse game is a play of transience, like that between netizens and censors or trendsetters and marketing researchers.

At the same time, perhaps, it is only within the confines of ephemerality that any kind of autonomy can exist. Anarchist writer Hakim Bey's theory of the Temporary Autonomous Zone (TAZ) thrives upon the temporality of the moment and the interstice spatiality of the fissures between hegemonic structures (1991). Initially written as a response to the "'anarchist dream' of free culture," Bey's ideas form an interesting juxtaposition with the "anarchist spirit" of China as described by novelist Yu Hua in his musings on the *shanzhai* phenomenon.[13] Returning to the four spaces presented in this paper, any illusions of anarchy, especially in the example of Womenjia, must be tempered by the tension between independence and autonomy. Does maintaining relative degrees of autonomy insinuate independence in the sense of an autarkic defining of borders? If all of these groups are working with *the community*, so to speak, the question must inevitably turn to the real distribution of so-called independence and the limits of these communities. That is to say, what are the communal parameters of a *we*? Where HomeShop and WooferTen make expressed attempts to address local publics with a mind to actively building new senses of community, it can be said they become ensnared in a socioeconomic framework that overwhelms such community, especially in the urban context. WooferTen's foundation as a government-supported entity implies a certain tenuousness with the alignments and/or frictions of belief and ideology, and the termination of their contract is, although not directly stated by the HKADC, contestably a political issue.

(In)Dependence, Industry, and Self-Organization

Figure 7: Passersby pause in front of HomeShop on the opening of the "Ten Thousand Item Treasury" public library. (photo courtesy of HomeShop)

For HomeShop, it may be very likely that the co-working framework it employed to sustain the space altered the field of locality first addressed by its being embedded within the *hutong*. A gated courtyard in its second Beixinqiao space and the steady stream of Chinese and foreign 文藝青年 *wenyi qingnian*, or "arty youth," displaced the initial focus, and HomeShop's members are very aware of being implicated in the "gentrification disco" of a creative Beijing. But like the discussion on the topic presented on HomeShop's blog or elsewhere, it is not enough to transplant urban concepts like gentrification and simply add Chinese characteristics (Eddy and Ren, 2013).

Womenjia, the most directly influenced by anarchist ideas of all the spaces, further complicates the question of resistance. Wuhan's notoriety as the birthplace of punk in China is no small impact on the space, but as Mai Dian laments, it is in reality the destabilizing noise of punk that has most ingeniously become a point of consensus between liberals and the state.[14] The gathering of massive crowds for the now common sight of music festivals in China is both a "collaboration between local governments' economics departments and the music industry," and it is one of the rare moments in which such large-scale public gatherings can be tolerated. In the widely circulated images of young Chinese punks in a mosh pit, "faces flush and passion rushes. In this sense, China has to a whole new level *shanzhai*'ed the West. In other words, it has learned and very well applied a 'flattening' government rule by hormones."[15] Womenjia's observation of such trends validates the further peripheralizing of itself in its own autonomy and explains the decreasing amount of visible activity in the space. In reflection, Mai asks, "How does 'anarchism' become popular in the tense moments of the anti-powers that be? How is combination or collaboration in the name of 'guerrilla' activity between creatives and those explorers who have realized the economic dynamics of mobility and transience in yet to be artistic spaces a new aesthetic? How could it become possible that we all, collectively, strayed away from production, or from the fact that we are part of the production of an artistic life?".[16]

Whether their manifestations in the making are directly informed by a mode of anarchism or not, it may be possible to say that the subjectivities of a Chinese anarchism have imposed itself on the daily life of these spaces. This is the same fractured time-space of Bey's TAZ subjectivity, in an era striated by the sedimented hegemonic practices of a state "simultaneously riddled with cracks and vacancies" (Bey, 1991). It is an inter-subjectivity founded upon rupture, not only from the dominant consensus but from the very notion that there is any "deeper objectivity exterior to the practices that bring it into being" (Mouffe, 2007). Theorist Chantal Mouffe's key point here is the lack of an outside to the *mise en abyme*; the only way through is via tactics, strategies, and practices, where community, independence, and exclusionism are agonistic struggles that layer over time. In this sense, autonomy and resistance are never ends in and of themselves, nor are they claims or monikers to which we must adhere. Instead, the "laboratory" of Womenjia and these other spaces exists as a series of experiments that examine, test, and re-test the validity of our existing sociopolitical concepts. Terms like "one country, two systems," "harmony," and "commons" may be used as forms of repression, but they may be just as much reactivated, and the flux in their understanding and usage is, of course, the same cat-and-mouse game that characterizes the ambivalence of resistance for each of the four spaces. The example of weak denial by Bartomeu Marí Ribas in the introduction of this text is a semantic resistance that inhabits the same micropolitically variable realm as Lijiang Studio, Womenjia, HomeShop, and WooferTen. In one sense, they are each exceptions to their immediate contexts, but what they offer are the slowly revealed processes of alter-rendering the sensible.

Figure 8: In September of 2011, Womenjia Youth Autonomy Lab hosted an activity to discuss the "Everyone's East Lake" project initiated as an open call for artistic interventions to raise awareness of questionable development along the perimeter of Wuhan's major body of water. Postcards were distributed with images of artworks and interventions from the project, co-organised by artists Li Juchuan and Li Yu. (photo courtesy of Womenjia Youth Autonomy Lab)

Sense of a Story: Forms of Representation

The international art world has welcomed the sensibilities of Jacques Rancière in recent years, with his texts on politics and aesthetics being widely circulated in tandem with the heightened interest in socially engaged and/or politically motivated art practices. It is Rancière's crucial idea of *le partage du sensible* that forms the link between aesthetics and politics, and it is the movement from "sense" to "sensibility" by which art and literature can expand and transform our perceptions of the world, forming propositions that maintain or challenge existing orders. Mouffe supplements this argument by placing such practices at the level of the psychic, whereby the political is precisely constituted by the "symbolic ordering of social relations" via human reciprocity, power, and aesthetic practices (2007). But indeed, is not the very tailoring of experiences and lifestyles by the creative industries the exact form of immaterial production that Žižek describes as "bio-political, the production of social life" (2012)?

What HomeShop, Lijiang Studio, Womenjia, and WooferTen have in common as opposed to more official institutions and emblematic forms of industry is a particular mode of self-organization and the self-representation that accompanies it. As mentioned previously in this essay, organization occurs here as an internally binding agent, and representation as an external one. What Mouffe places at the realm of the symbolic occurs just as much in the real space-time of living and working together. Even in the cases of WooferTen and Lijiang Studio, paper ties to the State in the form of municipal funding or tax-deductible status are forms of bureaucracy on paper that in many ways do not say much about the more spontaneous, ad-libbed manners of working that occur in the daily life of these spaces. And these narratives will vary continuously depending upon which aspect of organization is being examined. The modus operandi by which things get arranged—in the planning of a program, the sedimentation of old hierarchies during weekly meetings, or in the answers to Art & Labor activist Annie Shaw's simple yet challenging question, "Who does the dishes?"—all speak of organization at the levels of both the political and the aesthetic. They are political in that they determine, by a measure of internal to external differentiation, the bureaucratic distances of each of these spaces to other entities in society, namely government, funding bodies, and others in the local environment. A group becomes a group insofar as there is a relative consensus on a manner or attitude of affiliation that distinguishes it from others. Whether formalized or not, these affiliations and modes of collaboration also reveal the kind of power relations and inequalities that can, unfortunately, penetrate even an art-activist sphere.

An aesthetic understanding of an organization occurs in the tracing of activity towards the realm of representation. This occurs, of course, in the form, appearance, and structure of its activities as they are carried out and directly experienced, but it is also inherently tied to the politics of media and the increasing circulation of institutions and cultural activities via the internet and other forms of networked discourse. The organization of representation, as such, has become a highly sophisticated form of public relations that deals with the face of an organization in ways that are not necessarily tied to its actual, internal ways of working. This distance between the forms by which individuals and groups organize their practices as a daily actuality and the way this becomes manifested in a press release, artist talk, or Facebook page is an aesthetically mediated one, because it deals with language, the poetics of translation, and the creation of an image. It is narrative in the sense that there is a crucial time-space to be navigated in this distance. How one imagines an organization to be in the dreaming, brainstorming, and gathering of it evolves, for better or worse, in the actual realization of it over time. Whereas official institutions such as museums and commercial galleries are accountable to their constituents (e.g., the public and a board of trustees), self-initiated groups like Womenjia and HomeShop only need to qualify such relationships insofar as they feel personally interested and/or responsible. Representation, therefore, is not so much a matter of economic transparency, but of actively attempting to negotiate the boundaries between the internal and the external. To declare oneself as an "open platform" as these spaces often do is an uncertain extension of one's practices to an uncertain body of the public, beginning with the community and all the distraction that may include. As Mai Dian describes, "Once the gate is open, all kinds of disguised gods visit and declare their truths." [17]

Figure 9: In March 2011, members of the cycling circus troupe 2wheels4change opened Womenjia's gate to the public with their performance "金猴闹春 (*Jinhou Nao Chun*) The Golden Monkey's Spring Mischief." (photo courtesy of Womenjia Youth Autonomy Lab)

If this sounds frustrated, is the only recourse to return to a hermetic closing of the gate? How does one speak in a manner that gathers the kind of community one desires? And if such a community is honestly as open as one claims, how to represent the highly multiplicitous and divergent voices that gather? In the case of WooferTen, the professed dedication to the local community becomes a source of conflict when members disagree about the curation of activities. On one hand, "the local" stakes a limit on a geographically situated public, working with and for those in the immediate neighborhood of Yau Ma Tei, and within a realm of discourse pertinent to a Hong Kong identity. On the other hand, WooferTen's other emphasis upon art-activism opens a conceptual dialogue on the issue which is regional/global in nature, as per their East Asia Multitude series of symposia and invited residencies. What WooferTen's internal debate fails to resolve, then, is the pluralism inherent in the conception of the local, where a meaningful and engaged discourse with others addressing a similar set of issues, such as urban development and the threats of nuclear energy, can also comprise a contained locality.

In a similar logic, all four of the case studies presented in this analysis reveal a complex tension between understandings of local to non-local. Perhaps amplified by a "China versus the world" complex, all four groups can actually be regarded at least as Western-influenced explorations of collaborative practice (to what degree they may be labeled as *shanzhai* will be left to the reader), from the backgrounds of HomeShop's and Lijiang Studio's organizers to WooferTen's aesthetic referencing of space hijacking interventions abroad and Mai's conceptual nod to the European free-housing-for-all movement.[18] But to leave a reading of these practices at that, of course, does an immense disservice both to the work itself and the transcultural mediation of it. Referring again to Yu Hua's rescue of *shanzhai* culture, a copy, evolution or spin-off reveals that conceptual transplantations are never merely such; as much as critics deride copycatters' exploitation of originals (in the design sphere, for example), any attentiveness to the concept of origins, already diluted at best, cannot understand these initiatives without, admittedly, an acknowledgement of some sort of Western influence, but just as importantly, the entire other array of influences shaping what these spaces become. These are neither typologies, trends nor categorical phenomena. For each of the groups, there is a particularity of the local context, a mediated acculturation and a manner of representation. This is the subtle articulation of a sociopolitical position; it questions "how we work and participate in both the art world and public life," and it is a process of negotiation that makes up a movement rather than the solidity of a plaque and fixed form of representation; the two-way dialogue between organization and representation shifts in shape, in time, and in voice (Abu ElDahab et al., 2011).

Lijiang Studio's extremely detailed documentation of the mural project, while taking on the fixed format of a publication, is one such example by which the work may be actually more crucially revealed as processes of dialogue between organization and representation. The published volume, fittingly subtitled *Contemporary Art Episodes in Rural China*, includes personal reflections and negotiations between artists, curator, and locals, and reads more as a travelogue or series of short stories than an art catalogue (Li, 2012). At the time of its preparation, three years after the last mural project residency,

Li's return tour of the village revealed that several of the paintings had not survived weathering or their owners' aesthetic preferences, and perhaps just as well. A review of this project could never be summed up via a quantified value of each mural as a finished artwork, and it is just as problematic to turn the experiences between artist and farmer into a criterion for judgment. To do so would only allow a datafication of these relationships conducive to the stratified systems of non-stop production that Lijiang Studio tries to evade. Thus, it is exactly far away from the typical art-consuming audience that these works can only become narrated as a series of encounters and dialogues. Even where marks and comments from village householders are recorded and compared in the book, they serve more as nodes for discussion than fixed value judgments or systematized critique. He Simei, a teenaged villager who was vehemently opposed to the airplane Shanghai-based artist Liu Bin wanted to paint on their home, protested that her family would be ridiculed by others. But approval from He Simei's father and Liu's persistence led to the finished mural, which received a compliment from one of Simei's friends.

Figure 10: The airplane mural proposed by artist Liu Bin became a contested issue for the He family. (photo by Elaine W. Ho)

Liu interestingly observes Simei's tacit acceptance by the changes in her aesthetic sensibility; by the time he departs from Lijiang, she is experimenting with the cosmetic addition of "knickknack earrings, lipsticks and eyebrow pencils provided in a local snack package" (Liu, 2012: 81). After two years' time and having seen many similar paintings in the neighboring prefecture of Chuxiong, He Simei is finally able to convey to the curator her upgraded response: "It's not bad" (199).

This incident cannot reduce the artwork and these projects as a whole to a question of authenticity, where form would always equal content and a direction could be easily highlighted as the nostalgic return for something more "real."[19] In fact, it is that the scales are constantly shifting and should be continually challenged. When Mai serves as the primary representative of Womenjia, are we being misled by the "we" of Womenjia ("Womenjia" literally translates to "our house"), and what power structures come into play by this invitation? The same can be said of HomeShop, which began as the personal living space of a single artist experimenting with the seepage of a private space into the public sphere. But when the platform and the public space necessarily include participation from others, what questions of authorship emerge, and to what degree are individual subjectivities lost by old hierarchies and the sometimes tyranny of consensus? How much of the process of "commoning" is the use of a rhetorical language, and—as a manner of understanding pronouns, historicizing, and the simple expressing of desire—what does it return to an understanding of small-scale organization?

The Illusions of Autonomy: A Meta-Analysis

In the field of organization theory, the meta-theoretical analysis is one that pays close attention to these so-called intervals between form and representation. Sometimes noted as a "textual strategy" and at others a "rhetorical repertoire," institutional theorist Barbara Czarniawska makes an anachronistic leap to refer to the "styles of organization theory" (2003: 237-61). Where an initial dictionary comprehension of the term *style* points merely to formal qualities as elaborated in a sense of "elegance, refinement or excellence of manner, expression, form, or performance," the meta-analysis points towards those spaces between forms and their expression, revealing another pivot point between the politics and aesthetics of such artistic practice.[20] In this sense, the focus of the following analysis is not about the actual output of each organization, but the manner in which those outputs are voiced, and how that reflects back upon the nature of what is being said. Convoluting the discussion in this manner serves a twofold purpose: one, it deters any possible oversimplification by the comparative analysis of the four case studies presented; and two, it makes "stylistic" reference to the precise tensions that weigh upon the marriage of art and politics.

Bey refers to the livelihood of the Temporary Autonomous Zone as a clandestine operation carrying on underneath the nose of a state primarily concerned "with Simulation rather than substance" (1991). If so, WooferTen's highly visible participation on social media platforms, often to a great degree of simultaneity between activities and their documentation, concurs with their relationship to the state, in the sense that the highlighting of spectacle doubly serves as the kind of justification that funding bodies require. While this may certainly shift WooferTen outside of the TAZ, it is actually through WooferTen's consistently subversive use of media that we witness an extremely sophisticated rendition of "style as voice" in what Czarniawska very interestingly notes as both a political and an emotional sense (2003: 239). Although critics may point to the self-referentiality of these artists, highly complicit in the exhibition of art-activism without any real efficacy, the background of Hong Kong's relationship to the mainland must again be acknowledged in parallel to the very rise in mass-scale protests since 1997, often with artist-led participation.[21] With current political suspense mounting towards the constitutionally ambiguous movement towards universal suffrage for Hong Kong citizens in 2017, such a consideration of the voice is political, emotional, *and* literal.[22] Their performances in the theatre of representation are imperative because it is representation itself which is at stake. WooferTen's annual costumed bicycle procession on June Fourth may be a simple and crude upholding of a visual mimesis, but its power lies in the collaborative jolting of collective memory in public space where it does not exist elsewhere.

(In)Dependence, Industry, and Self-Organization

Figure 11: Participants gather outside of WooferTen (*at left*) before the annual bicycle procession from Yau Ma Tei to Victoria Park on June 4th, 2013. (photo by Elaine W. Ho)

Rancière's discussion on the dual autonomy and heteronomy of the artwork is pertinent here, for it is not the mode of presentation (individuals dressed as mainland students from the 1980s) which makes the work, but the particular undoing between individuals, "in that zone before representative sequences, where other modes of presentation, individuation, and connection operate" (2004: 148). Where such an experience has been maligned by cynics as the useless exception of a 364-day annually numbed population, perhaps it is precisely the persistence of WooferTen's activities within a dueling two systems that should be regarded as the starting point of evaluation.

Such a direct gathering of the voices of dissent is, of course, not a possibility for the other three groups based inside the mainland, and the dispersed appearances of Lijiang Studio, Womenjia, and HomeShop across the country can be likened to micro-zones of fortitude amidst the hegemony of market socialist-approved cultural industries in the PRC. The ambivalence of their modes of representation (Womenjia stays largely off the radar of the Internet and social media; Lijiang Studio's website was inactive for over one year, since hacked by spammers; and HomeShop very often conscientiously employs an ambiguous artistic language in its self-representation) is the crux of a dialectic

[209]

between autonomy and heteronomy, art and politics. As singular efforts that weave through the loopholes in the blanket of overhead systemic issues, they are forms of production that insist upon maintaining a degree of autonomy while being simultaneously cognizant of the ineffability of certain ineludible frameworks. Narratively speaking, this manner of operating communicates out of error and in difference. Womenjia and HomeShop speak from the place of idealism, yet they acknowledge their explorations serve in no way as utopian models. Lijiang Studio is perhaps more modest in its ambitions to make "art that is as interesting to the visiting artist as it is to local people as it is to us," but this relaxed parlance is carried out by the same efforts to expand and disturb existing forms of discourse in its community (Li, 2012: ii). There is a lapse traversed when, as artist-writer John Miller says, "Direct experience migrates into representation," (2012: 342) and the style that emerges amidst these exchanges happens as change in relation; it is the active process of autonomy in discussion with heteronomy. In other words, it is about understanding WooferTen, Lijiang Studio, HomeShop, and Womenjia in terms of the ongoing relations between the work itself and the means of engaging it. What Czarniawska describes for the work of literature applies equally to the work of art: "Style is the writer's awareness of being engaged in writing, incorporated into the text itself" (2003: 240). If these groups in some ways fit categorical bills for TAZ or "self-organized," "alternative" forms of "social practice," it must also be acknowledged that each of these terms cannot simply be analyzed from a global perspective. And where art historian Maibritt Borgen recognizes a more complicated reality of self-organization by its inner and outer forms, the emphasis in this reading goes further to meander parallel to the movements between, around, and through these skins (2013: 37-49). To ambiguously say, "they have style" refers to Rancière's theoretical torus, whereby

> Literature is produced by making itself invisible, by combining the molecular music of affects and free percepts with the molar schemes of representation. The literary power of style thus becomes, in the final analysis, identical with the art of the Aristotelian mimetician, who had to know how to hide himself in his work. Here it is literature itself that hides its labor by accomplishing it, that makes indifferent the difference that

results from the principle of indifference, from the principle of non-preference (Rancière, 2004: 151).

The conclusion refers back to the openness of these considerations, despite the distinctions made to pinpoint them. On one hand, we can consider all four of the case study initiatives as a form of social practice in China. But they are considered here post-mortem, or with an undefined experience of what the work actually is. The politics of this practice, can perhaps only be lived through experience, rather than issued as a statement. They are creative practices mapped only perhaps as the telling of an infinitely detailed story, where the dialogical practice of stirring modes of community is intricately linked with their representation and the politics of their aesthetics. In this sense, the "story" is not simply a manifesto or the moral we are left with at the end; it is the active, ongoing process of the telling and re-telling of stories as the sometimes fluent and sometimes incomprehensible narration of allegories— sometimes also jokes. Where the creative industries fall prey to the market is the inevitable valuation that occurs amongst all the different "stylizations," but the counter-punchline here is a passing of time that refuses the equivalency of meaning and value. As even the anarchist Mai Dian admits, the unprecedented wealth of resources (e.g., labor and private finance) to support the circulation of artistic production in China today means that its participants are almost always both resistors *and* clients. This falls in line with Diedrich Diederichsen's analysis of artistic *mehrwert*, which "yokes together two different things: on the one hand, the conceptual accreditation of artistic movements that abstract concrete objects and introduce the resulting abstractions into critical projects; on the other, the instrumentalization of these abstractions by an abbreviating culture of communication" (2008: 27). The implication is a heady realization: we are at once the exploited and the exploiters, the characters in the story as well as the storyteller.

Perhaps these contradictions are just more of the clichés to be added to the list of Chinese characteristics. At the same time, it is not the aim of this paper to color the map that is independent arts practice in China today. WooferTen, Lijiang Studio, HomeShop, and Womenjia are four examples, but they are depicted as narrative springboards for thought rather than offerings to dissect a market-socialist phenomenon. The spontaneous forms of agency and innovation to have emerged from these spaces must be emphasized inasmuch as they can be experienced. Hardly a call to the commodification of experience, these spaces may just as well transform and/or dissolve into other trajectories, associations, and stories. In the same manner that they have come about in rupture and displacement from an existing set of political and aesthetic arrangements, they disrupt their own modes of working in search of other methods. Whether this entails future forms of rupture in physical space, the stakes of discourse and autonomy within art, or an "aesthetic rearrangement" of social hierarchies has yet to be seen. And that is to say, it explores politics and aesthetics as that which has yet to be mapped, where industry points to processes rather than products and a description portrays what it is never capable of fully depicting. There is a stake on unnamed potentialities and the fallout that ensues from its misinterpretation. And Ribas may very well be aware of that.

Figure 12: The WooferTen neighbour known to the community by the name "Fred Ma" distributes free food to visitors. (photo by Elaine W. Ho)

Notes

1. Recorded on the occasion of the Multitude Art Prize Discourse Series held at Ullens Center for Contemporary Art in Beijing, April 28, 2013. For more information and full documentation from the conference proceedings, please visit *www.multitudefoundation.org/conference.html*.
2. In his keynote speech to the 17th National Congress of the Communist Party of China (CPC) in October of 2007, President Hu Jintao stressed the importance of Chinese culture and the cultural industries "as part of the soft power of our country to better guarantee the people's basic cultural rights and interests."
3. The term "relational repression" refers to a common technique of "soft" repression that combines state and personal social relations to diffuse dissent and protest. See further information in Yanhua and O'Brien (2013: 533-52.)
4. Mai, e-mail message to author, 7 October 2013.
5. Brown, e-mail message to author, 9 September 2013.
6. For further reading about visions for the new socialist countryside and the New Rural Reconstruction Movement, please see texts by Renmin University School of Agricultural Economics and Rural Development dean Wen Tiejun; an overview in English can be found at "Building a New Socialist Countryside": *http://www.china.org.cn/english/zhuanti/country/159776.htm*.
7. "Foreign-supported platform" refers to Lijiang Studio's status as a 501c3 not-for-profit foundation in the United States and, for the purpose of institutional equivalency in China (a requirement for American non-profit organizations to be able to transfer funds abroad), the Lijiang Association for Cultural Research and Development in the city of Lijiang.

8. See a detailed contemporary history of politically-charged Hong Kong art in Lau (2012).
9. Under the terms of agreement with the Hong Kong Arts Development Council, WooferTen should have evacuated 404 Shanghai Street in the autumn of 2013, but at the time of this writing in the winter of 2014-2015, members have yet to clear from the space.
10. I will begin this introduction with the disclosure that my position as initiator of HomeShop will alter the subjective positioning of it relative to that of the other spaces presented in this paper, but I hope that will not diminish the value of its bearings within the larger discussion of artist-run spaces in China. In some respects, the later discussion on narrative forms of representation is also relevant to my role as author of this reading. More information about HomeShop and some of its projects can be found at *www.homeshop.org.cn*.
11. It is interesting to note that the term "alternative" to describe art spaces was inadvertently declared dead already in the West by 1985, per artist-writer Julie Ault's analysis of the New York art scene in her book *Alternative Art New York, 1965-1985*. Why it would be picked up again by Chinese art circles in the 2000s is partially cliché, though partially also a discrepancy in translation.
12. Ironically, Žižek credits the Chinese socialist market economists with being "the most efficient managers of capitalism because their historical enmity towards the bourgeoisie as a class perfectly fits the tendency of today's capitalism to become a managerial capitalism without a bourgeoisie—in both cases, as Stalin put it long ago, 'cadres decide everything.'"
13. "It would not be going too far to say that 'copycat' (*shanzhai*) has more of an anarchist spirit than any other word in the contemporary Chinese language." Yu (2012: 181-182).
14. Mai was the editor of the independently produced leftist zine 冲撞 (*Chongzhuang*) *Chaos* (2001-2005) and member of several punk bands in Wuhan, the most recent of which is called 犯罪想法 (*Fanzui Xiangfa*), meaning "Criminal Thoughts."
15. Mai, e-mail message to author, 7 October 2013.
16. Mai, *ibid*.

17. Mai, e-mail message to author, 6 October 2013.
18. Lijiang Studio was established by American Jay Brown, and HomeShop was initiated by Chinese-American Elaine W. Ho, co-organized since 2011 by artists Michael Eddy (Canada), Fotini Lazaridou-Hatzigoga (Greece), and Emi Uemura (Japan), and theorist/cultural workers Ouyang Xiao (China/USA), Twist Qu (China) and Cici Wang (China).
19. As an aside, it is also interesting to note that the valorizing idea of authenticity is much more so a Western concept, and as an example, the prevalence of the *shanzhai* phenomenon in China bequeaths authenticity passionately in favor of what Yu Hua describes as "revolutionary action initiated by the weak against the strong." This is evidenced by everything from *shanzhai* entrepreneurialism by small-scale production (everything from electronic goods to luxury products and toilet paper) to *shanzhai* pop culture that, thanks to the internet, makes stars of its protagonists.
20. *New Shorter Oxford English Dictionary* (Oxford: Oxford University Press, 1993) in Czarniawska (2003: 239).
21. One of the most visible examples of such show without "real efficacy" was the highly media-focused spotlight of Choi Yuen Village during the 2009-10 protests against the development of the Guangzhou-Hong Kong high-speed rail. While the actions at that time catalyzed many of the relationships and strategies important for Hong Kong's artists and activists since then, it failed to achieve the stated goal of saving the village from demolition.
22. Hong Kong's Chief Executive is currently elected by an Election Committee composed of voters from the functional constituencies, religious organizations, and municipal and central government bodies, though the increasing infusion of a pro-Beijing electorate has led to widespread controversy over the possible reform hinted at in Article 45 of Hong Kong's Basic Law, which claims universal suffrage for the Chief Executive as an ultimate goal. Wikipedia contributors, "Politics of Hong Kong (section Universal suffrage)," *Wikipedia, The Free Encyclopedia*, http://en.wikipedia.org/wiki/Politics_of_Hong_Kong#Universal_suffrage (accessed 18 October 2013).

Bibliography

Abu ElDahab, Mai, Binna Choi, and Emily Pethick, eds. *Circular Facts.* Berlin: Sternberg Press, 2011.

Bey, Hakim. "T.A.Z.: The Temporary Autonomous Zone, Ontological Anarchy, Poetic Terrorism." *The Hermetic Library.* 1991. www.hermetic.com (accessed 5 October 2013).

Borgen, Maibritt. "The Inner and Outer Form of Self-Organization." In *Self-Organized*, edited by Stine Herbert and Anne Szefer Karlsen, 37-49. London: Open Editions, 2013.

Ching, Chin-wai Luke. Research interview by Alice Ko. Hong Kong, 8 May 2013.

— Press Release and Curatorial Statement of the exhibition *Hong Kong Needs "Hijacking."* Hong Kong: White Tube Gallery of the Hong Kong Art School. Hong Kong Arts Centre, October 2008.

Czarniawska, Barbara. "The Styles and Stylists of Organization Theory." In *The Oxford Handbook of Organization Theory: Meta-Theoretical Perspectives*, edited by Christian Knudsen and Haridimos Tsoukas, 237-61. Oxford: Oxford University Press, 2003.

Diederichsen, Diedrich. *On (Surplus) Value in Art.* Berlin: Sternberg Press, 2008.

Eddy, Michael, and Julie Ren. "Gentrification Disco, Vol. 6: On the Problem of Transplantation," *HomeShop blog.* 20 March 2013 www.homeshopbeijing.org/blog/?p=4852.

Kao, Jun-honn. "巨人屁眼裡的一根針: 武漢我們家青年自治實驗室 "Needle in the Ass of a Giant: Womenjia Youth Autonomy Lab." 藝術觀點 *Art Critique of Taiwan* 53, (6 July 2013) 134-142.

Lau, Kinwah Jaspar. "Introduction of 'WooferTen.'" In *Creating Spaces: Post-Alternative Spaces in Asia*, edited by Lu Peiyi. Taipei: Garden City Publishers, 2011.

— "Politics of a Bio: Hong Kong Art from Dissemination to Usage." In *Hong Kong Eye: Contemporary Hong Kong Art*. Milan: Skira Editore S.p.A, 2012.

Li, Lisha Lisa, ed. *Lijiang Studio Mural Stories: Contemporary Art Episodes in Rural China*. Beijing: 知识产权出版社 Intellectual Property Rights Publishers, 2012.

Lijiang Studio. "Brief intro to 'Mural Project of New Countryside Laboratory' of Lijiang Studio." www.lijiangstudio.org/web/Studio.jsp?RecordID= 20090528135722&FunctionID=20050429144915&FirstMenuID= 20050429144555 (accessed 30 September 2013)

Lu, Yinghua Carol. Press Release of *Little Movements: Self-Practice in Contemporary Art*. Shenzhen: OCT Contemporary Art Terminal, 10 September - 10 November 2011.

Mai, Dian. "Exodus," *Wear* Journal 3 (2012): 66-7.

— "Resistances within a Public Ruin." In *Creative Space: Art and Spatial Resistance in East Asia*, edited by Hui Yuk. Hong Kong: Roundtable Synergy Books, 2014.

Mouffe, Chantal. "Artistic Activism and Agonistic Spaces." *Art & Research: A Journal of Ideas, Contexts and Methods*, Vol. 1, No. 2 (Summer 2007).

Miller, John. *The Ruin of Exchange*. Zürich: JRP-Ringier, 2012.

Rancière, Jacques. *The Flesh of Words: The Politics of Writing*. Stanford: Stanford University Press, 2004.

Yanhua, Deng, and Kevin J. O'Brien, "Relational Repression in China: Using Social Ties to Demobilize Protesters." *The China Quarterly*, Vol. 215 (24 July 2013) 533-52.

Yu, Hua. *China in Ten Words*. Translated by Allan H. Barr. New York: Anchor Books, 2012.

Žižek, Slavoj. "The Revolt of the Salaried Bourgeoisie." *London Review of Books* Vol. 34 no. 2. (26 January 2012) 19-22.

CHAPTER SEVEN:

On the Finding, Producing, and Losing of Creative Space: Hong Kong's Hidden Agenda
Tobias Zuser

Over the past decade, Hong Kong has positioned itself as the foremost arts hub in Asia, drawing upon large-scale investment to bundle cultural forces in the reclaimed West Kowloon Cultural District. However, due to changes in economic and urban structure, less visible creative clusters have independently taken root in the industrial areas of the city. Although the visual arts scene in Fo Tan has received international attention, this research is concerned with music spaces in Kwun Tong, using the illegal venue Hidden Agenda as a case study to discuss spatial practices and policy discourses in the city. Aiming to understand the prospects and remainders of cultural diversity in Hong Kong, this chapter is divided into three parts: the general development of autonomous cultural space in the city (*finding space*), a spatial analysis of the industrial area in Kwun Tong (*producing space*), and a concluding discussion on the imminent precedence for the future cultural planning of Hong Kong (*losing space*). In order to connect these various topics, this study takes on an interdisciplinary approach across cultural and urban studies. An adaptation of Lefebvre's conceptual triad of perceived, conceived, and lived space is used to understand the production, appropriation, and elimination of cultural space in the face of urban redevelopment projects in Hong Kong. Eventually, this chapter will examine the cultural, social, and economic sustainability of Kwun Tong as one of the most thriving areas for creativity in the city, while using Fo Tan as an intra-urban reference to emphasize its particularities.

For more than a decade, Hong Kong has condensed its strategies for cultural development to a single project and placed collective hopes and expectations on the shoulders of the West Kowloon Cultural District (WKCD). However, as of 2013, this 40-hectare area of reclaimed land at the northwestern shore of Victoria Harbour, which has been reserved for the construction of some of the most comprehensive arts venues in Asia, has not become any more than a representation of what could, should, and might be. Nevertheless, the continuous habit of the city to speak in the future tense seems to have eventually created a neglect of what—in the meantime—has already been materialized somewhere else by someone else. Lily Kong (2012: 182), a leading scholar of creative economy in the region, has referred to Fo Tan, one of few organically evolved creative areas in Hong Kong, as "an unlikely arts cluster in an unlikely city" that emerged at a time when the WKCD was still an empty piece of land. More than twelve years later, the former remains one of the most vital spaces in Hong Kong for producing visual art, whereas the productivity of the latter is still deferred by disputes over positioning, scale, and subsidies.

Surprisingly, articles that discuss the creative potential and development of Hong Kong are often determined by their anticipation for the WKCD or the analysis of a handful of culturally reframed heritage sites.[1] However, the inherent cultural resources for creative economies seem to be rarely picked up in the local context, as vocally proclaimed by the commonly known mainstream discourses that affect cultural policies worldwide (Landry, 2000; Florida, 2002). With the rise of Hong Kong as one of the leading art auction markets in the world because of its geopolitical advantage in the Asian region and its mediating role for China, the local visual arts scene, which is partly hidden away in the industrial district of suburban Fo Tan, has truly gained more attention both domestically and internationally. However, a meaningful connection is hardly ever built on the ways these specific spatial practices are embedded in and related to urban cultural planning.

Nevertheless, although certain critical articles have been published in recent years, such as those related to the sustainability of Fo Tan as an organically evolved cultural cluster (Kong, 2012) and the commercialization and ahistorical appropriation of the colonial heritage of Hong Kong (Ku, 2010), the actual significance of such "unlikely arts clusters" and their (dis) placement within the politics of creativity in Hong Kong often remain unclear.

This chapter aims to investigate these missing links by focusing on both the socio-spatial construction and sustainability of a creative space[2] located in the midst of an industrial area in Hong Kong. Although the awareness of clusters such as Fo Tan has steadily increased since the upscaling of the annual visual art fair,[3] this study intends to redirect the focus on the local independent music scene and its preference for Kwun Tong, a district in the east of the Kowloon Peninsula and, interestingly, very much the geographical counter-pole of the aforementioned WKCD. However, this shift in the research object is also informed by an actual state of urgency: the government has targeted most of the industrial buildings in Hong Kong for commercial redevelopment, and Kwun Tong will soon set a precedent for either the sustainability or destruction of the autonomous creative spaces in the city.

This chapter is separated into three parts. The first section will start with a brief historical and contextual outline of the emergence of a spatial dimension within the cultural policy discourses of the city and describe how eventual recognition has created a symbolic case for the decentralization of creative spaces. It will then further investigate the attributes and values industrial buildings provide for the continuous production of culture in a city with some of the most expensive property rental prices in the world and how these conditions have been challenged by recent redevelopment plans. This finally leads to the introduction of the Hidden Agenda, an illegal music venue that will function as a case study to illustrate the material conditions and cultural significance of such spaces.

The second part of the chapter will consist of a detailed account of the spatial practices and representations that Kwun Tong, through its various perceptions and conceptualizations, produces and is constantly produced by, drawing on the theorization of space as a social product (Lefebvre, 1991 [1974]; Harvey, 1990). Given the interdisciplinary nature of this publication, the analysis is inspired by the "ground-truthing" methodological use of Lefebvre's conceptual triad as suggested by Jana Carp (2008) to offer a more accurate account of the complexities involved in the production of creative clusters in Hong Kong.

Finally, in the third part, these findings will be used to assess the cultural, social, and economic sustainability of the Hidden Agenda using the aforementioned existing research by Kong (2012) as a reference point. Kong's study is used to understand the distinctive features and conditions of the music cluster in Kwun Tong compared to the organically evolved visual arts scene in the industrial area of Fo Tan. All analyses proposed in this chapter are based on observations and data collection during fieldwork in Kwun Tong in 2013 for an ongoing research project about Hong Kong's cultural policy.

Part I: How to Find Space

In his first policy address in 2013, the newly appointed chief executive of the Hong Kong Special Administrative Region (or Hong Kong SAR), commonly referred to as CY Leung, articulated the term "space" with three different domains: housing, land supply, and arts (Leung, 2013). The latter reference is stipulated by the cultural vision to "give young artists and new arts groups more room for development," and concrete strategies, such as the provision of creative space in an industrial building in Wong Chuk Hang (Leung, 2013: 66). In fact, this instance was the first time that the need for creative space was recognized as a policy concern by an acting chief executive.

However, the demand for arts spaces in the city has taken a prominent stand in cultural policy debates as early as 1998, when more than twenty arts groups occupied the abandoned Government Supply Depot in Oil Street on Hong Kong Island (Cartier, 2008). The artists were offered a temporary lease, but had to leave the premises in 2000 because the site has been reserved for redevelopment. Following negotiations between the government and the artists about the necessity of such creative spaces under the S.O.S. Save Oil Street campaign, most of the arts groups were provided with an alternative location at the smaller Cattle Depot Artist Village in Kowloon East, which was previously used as a slaughterhouse. The establishment of this cultural cluster, which remained under the supervision of the government, coincided with the emergence of a less visible arts scene in the industrial areas of the city left from a time when Hong Kong was one of the biggest producers of light consumer goods in Asia.

In the years after the Second World War, Hong Kong received roughly 1.3 million immigrants, and this sudden excess of labor boosted the economic development of the region in the years thereafter (Leung and Soyez, 2009). While during the 1950s and 1960s the industry was mainly dominated by textile entrepreneurs who left Shanghai after the establishment of the People's Republic of China, Hong Kong had become the world's largest toy producer by

1979 (Young, 1992), constantly requiring new production sites close to logistic nodal points such as harbors and airports. However, this situation changed drastically under the reform policies of Deng Xiaoping.

An increasing number of buildings in the industrial areas of the city became redundant mainly because Hong Kong's manufacturing industry was relocated to mainland China, where the production of the same goods was significantly cheaper. The Asian financial crisis in the late 1990s also aggravated this situation. Following similar transformations of deteriorating factories in port cities all around the world, local artists soon recognized the opportunity to rent flat-sized units at reasonable prices to open up their own private studios or rehearsal rooms; Fo Tan, Kwun Tong, and Chai Wan became the most popular districts for these endeavors (Cartier, 2008). However, according to the stringent town planning ordinance in Hong Kong, the permitted use of spaces in an industrial building is restricted by the zone category in which the premise is located. Thus, these peripheral clusters are often barred from commercial or residential gentrification and are literally out of sight of the cultural displays in Hong Kong, with the positive side effect that rental costs in these areas tend to be far below the city average. But this had been the case only until recently.

Given that further land reclamation in the Victoria Harbour area is prohibited, the opportunities for new real estate projects on Hong Kong Island are rare. The limited supply of office space within the Central Business District (CBD) has led to skyrocketing rental prices, which displaces an increasing number of small and medium enterprises from the city center. Therefore, the unused potential of seemingly redundant industrial areas has received new attention.

Currently, Hong Kong has around 1,400 industrial buildings, the majority of which are so-called multi-story flatted factories. As one building usually houses different manufacturers or warehouses under the same roof, ownership is often fragmented, which complicates the transaction of properties in these areas. Thus, in April 2010 the government introduced measures to encourage the revitalization of underutilized industrial buildings by offering favorable conditions for acquiring, converting, and redeveloping such premises. These measures led to a significant revaluation of industrial buildings and a sharp increase in rent for their tenants. According to an investment company, no other asset class in Hong Kong has been valorizing as significantly for the last five years (Bloomberg, 2012).

The measures also had a major effect on the nurturing conditions of arts spaces in industrial areas. As a reaction to the increasing concerns expressed by affected artists and organizations, the Arts Development Council (ADC) conducted a survey on the "status of industrial buildings for arts activities and future demand" in November 2010. The report on the survey findings showed that 64.7 percent of arts practitioners in industrial buildings were relatively young (between ages 20 and 39) and mainly engaged in music and visual arts (each around 37 percent). In its conclusion, the ADC confirmed that most of these artists have been adversely affected by the revitalization measures, and concluded its report with concrete policy recommendations for the government (2010). Since then, newly formed organizations, such as the Factory Artists Concern Group, have attempted to increase both public and political awareness for the creative use of industrial buildings.

Eventually, as mentioned earlier, the latest policy address in 2013 underlined the necessity for creative space, and referenced the appropriation of an industrial building in Wong Chuk Hang in the southern part of Hong Kong Island. Since Hong Kong has increased in significance as a global hub for the visual arts market in Asia, this area has become a popular base for art galleries that rate the need for a large exhibition space over a location in the central business district (SCMP, 2013). Simultaneously, the ascendancy of the annual art fair, the increasing revenues of auction houses, and the pervasive branding campaign of the future "M+" museum for contemporary art in the WKCD have also increased the awareness for and engagement with the local visual arts scene. In January 2013, more than thirty thousand people visited the annual open studio event in Fo Tan, which is equivalent to a fivefold increase of visitors since 2007.

However, the dominance of the visual arts scene in Hong Kong has also marginalized organically evolved clusters where other art forms prevail and where the symbolic, economic, and cultural values attributed to this scene seem to be less developed. Kwun Tong and Ngau Tau Kok, for instance, are known for their concentration of rehearsal rooms and, according to the ADC survey (2010), these districts together accommodate one third (32.3 percent) of all industrial arts spaces in the city, outpacing other areas such as Fo Tan (20.5 percent) and San Po Kong (10.6 percent).

The Hidden Agenda is a live music venue located in the industrial area of Ngau Tau Kok, a sub-district of Kwun Tong. Originally a privately rented band rehearsal room, it was converted into a space for public performances by its tenants back in 2009. Providing a platform for local, independent music bands, often based in industrial buildings nearby, the Hidden Agenda also organizes concerts by foreign musicians on a regular basis. The majority of the artists can be associated with alternative genres, such as punk, hardcore, indie pop/rock, reggae, R&B, and electronic dance music. Depending on the popularity of the performer, a regular show can attract between thirty and three hundred visitors, the latter also being the maximum capacity of the venue. Currently, the Hidden Agenda stages around six to ten live shows every month. In 2012, more than six hundred bands were given the opportunity to perform. Before the Hidden Agenda opened its doors to the public, similar concerts by local bands were often restricted to non-profit youth centers.

Venues with comparable capacities can be found in Sheung Wan, Central, and Wan Chai. However, they are primarily run under the regulations for bars, cafés, restaurants, clubs, or galleries, and offer live music only occasionally. The Hidden Agenda hardly perceives them as competitors, and vice versa. As regular businesses, they depend on the daily spending of customers to cover the high operational costs in these districts and are, therefore, often more selective in creating programs for their events. Given that they are mainly located in the city center, they are also subject to stricter enforcement of permits and noise nuisance laws. Popular concerts that attract larger audiences are usually staged in specialized event locations, such as the Kowloon Bay International Trade & Exhibition Centre (KITEC) or the AsiaWorld-Expo.

However, the near future of the Hidden Agenda seems to be uncertain. Over the last four years, it has been challenged by measures for revitalizing industrial buildings and continuous efforts by authorities to render the space illegal. Finally, in 2010 it had to leave its original space because the proprietor sold the premises. They moved to a new location, where the new landlord allegedly terminated the contract after the Lands Department considered taking legal action against them due to "unauthorized use." The current space of the Hidden Agenda, located in Tai Yip Street, is around four thousand square feet (or 372 square meters) and is rented for around HK$25,000 per month, which equals an increase of 150 percent compared to its previous location. The operators do not expect to survive in the long term and think that a closure will be inevitable by the end of the year. At the same time, the feasibility of proper legalization remains unlikely for several reasons. First, the statutory requirements to register the venue as a public place for entertainment have been tightened by the different town planning zones and the permitted use of industrial buildings within the area. The Hidden Agenda could apply for a waiver at the Lands Department by paying a fee of HK$120,000. However, this investment would not guarantee the approval of the waiver. Second, given that entrance fees are charged for most of the shows, the operators would also need to apply for an entertainment license from the Food and Environmental Hygiene Department (FEHD), which also imposes fire safety regulations that can hardly be met by building standards in industrial areas. Third, selling alcoholic drinks is also prohibited unless the FEHD has granted a liquor license. Despite all these impediments and regular visits by authorities, the Hidden Agenda has managed to continue its operation, mainly by using legal tools to appeal or at least delay decisions by different government departments. Together with Revitalisation Internalise Partnership (R.I.P.), a concern group that frequently coordinates campaigns and political rallies, the Hidden Agenda also continues to document and publicize its ongoing challenges with the law, with an understanding of its role as a representative for the larger arts community in Kwun Tong.

Although no official figures are available, several insiders estimate that on average each of the three hundred industrial buildings in Kwun Tong (including Ngau Tau Kok) accommodates at least three bands, which would result in a total number of roughly one thousand active music groups. Compared to Fo Tan—which emerged as a cluster for visual arts mainly due to its proximity to the Fine Arts Department of the Chinese University of Hong Kong—the dominance of music in Kwun Tong must also be seen in a wider socio-economic context. The political district of Kwun Tong, which also includes Ngau Tau Kok and Kowloon Bay, is the most densely populated district of Hong Kong, with more than 52,000 people per square kilometer, and houses around 580,000 residents in total. It is also one of the poorest districts, with more than half of its population living in public housing estates. Nevertheless, the population is currently estimated to grow rapidly over the coming years, reaching 900,000 by 2019 (Finance Committee, 2013).

Therefore, given its large, underutilized industrial area, Kwun Tong has been a major target for large-scale redevelopment projects in recent years. Since 1998, Sun Hung Kai Properties have been pursuing their main project in that area, Millennium City, encompassing five modern office buildings including the APM shopping mall, which was completed in 2005. In addition, the Kwun Tong Town Centre Project was started in 2013 after several years of consultation, and is said to be the largest single project ever undertaken by the Urban Renewal Authority (URA), affecting more than five thousand residents and five hundred shops. According to current zoning plans, Kwun Tong will also be significantly affected by the Kai Tak Development Project under the Civil Engineering and Development Department, which aims to convert the land used by the former airport of Hong Kong into a tourism and entertainment hub (Leung, 2013).

Part II: How to Produce Space

Based on the contextual framework laid out above, the second part of this chapter will analyze the specific spatial practices that have offered both the conditions and distinctions for the production of creative space in Kwun Tong. Marxist philosopher Henri Lefebvre (1991 [1974]) suggested an intertwined model of duality for his conceptual triad of space that consists of the three pairs (spatial practice/perceived space, representation of space/conceived space, and representational space/lived space), which will be further explained below. Jana Carp attempted to apply these concepts directly to issues of urban planning. Instead of using these terms as interchangeable ambiguities similarly to most scholars, Carp argues for an interpretation that stresses the dialectic interrelationship of these dualities (2008: 131). Such interpretation provides a simultaneous understanding of space as both social construct and process, following Lefebvre's instruction not to treat it merely as an abstract model (1991 [1974]: 40).[4] Therefore, this analysis should not only discuss how creative spaces are produced by the existing materiality (e.g., the actual building, environment, district in which the space is located), but also how different agents inform this production through their own embodied, mental, and social experience (Carp, 2008: 131). Given the imminent redevelopment of the industrial area in Kwun Tong, Carp's pragmatic approach can help "recognize divergent, incommensurate experiences that are directly related to the physical characteristics where intervention is being considered" (136). Official representations often communicate one homogenous "true space." Conversely, Lefebvre's "truth of space" concept argues for the recognition of multiple experiences, which, although often completely different from one another, all remain true simultaneously.

However, compared to Carp, who uses this method to conciliate across different professions (or "truths") involved in the actual planning of development projects, this study is limited to a selective analysis of two opposed conceptualizations of space that are informed by authorities and developers on the one hand and current users (i.e., artists and operators of creative venues) on the other.[5] Rather than offering concrete suggestions to reconcile different ideologies, the dichotomy will allow the articulation of the findings with questions of cultural policy in general and sustainability in particular.

As suggested earlier, this chapter turns its attention to the industrial area of Ngau Tau Kok, which is embedded in the larger political district of Kwun Tong in the east of Kowloon, but at the same time is separated from its surroundings by "natural" borders: to the southwest by a flyover highway construction (Kwun Tong By-pass) and the harbor (Kwun Tong Typhoon Shelter), and to the northeast by busy arterial streets (Kai Fuk Road and Kwun Tong Road) and the viaducts of the Kwun Tong subway line. This geographic constellation determines the physical framework in and through which the perception and practices of each user take shape. Indeed, the first layer of Lefebvre's conceptual triad—spatial practices/perceived space—is merely concerned with an empirical evaluation of the bodily experience of space and how it informs the patterns and behaviors of our daily routines and rituals. In other words, the production of space starts from the way we move within physical space while accommodating its sound, texture, smell, and shape through all of our senses. Of course, each perception differs from one person to another, but they drastically depict the discrepancies between internal users and external planners. In reality, the embodied experiences of actual locals are often overruled by non-local professionals who claim to provide a more sophisticated understanding of the space based on their expertise (Carp, 2008: 132-4).

Obviously, different means of transportation shape the routes used by regular users of the industrial area. For most people, Kwun Tong is easily accessible by subway. In fact, the Kwun Tong line was the first completed section of the comprehensive Mass Transit Railway (MTR) network in Hong Kong and has been operating continuously since 1979. Nowadays, Kwun Tong can be reached by MTR within twenty minutes from the commercial focal point of Mong Kok, Kowloon. Additionally, numerous established bus routes connect one of the most densely populated districts with all other parts of Hong Kong.[6]

Accessing the industrial area by foot from the nearby Ngau Tau Kok subway station will first lead the visitor through a lengthy underpass that ends at the northern edge of Lai Yip Street. From there, a short walk of about 450 meters leads to the Hidden Agenda,[7] which, in the following detailed description, serves as an illustration of how the physical characteristics of the industrial area inform a distinct set of spatial practices.

The front sides of most premises along the way consist of wide doorways for the loading and unloading of goods. This pattern is only occasionally interrupted by private stores, such as car repair shops or wholesalers; and in contrast to commercial or residential areas in Hong Kong, the prohibition of retail and entertainment businesses keeps the private consumer or *flâneur* at bay. Consequently, sidewalks are significantly less frequented, and the generally fast pace of the city appears slower than usual. The only common spaces in this area are a handful of eateries and convenience stores that are widely dispersed, and spatiotemporal routes and daily rituals of different users—regardless of whether they are workers, artists, or visitors—eventually cross each other here.

However, the most apparent *sensory* transformation along this particular way takes place after the transition from day to night. Although some factories and stores operate all day, traffic almost ends during evening hours. As a result, sidewalks and pedestrian crossings are increasingly perceived as superfluous, and nocturnal users appropriate the generous space of the streets to modify the efficiency of their own routes. Meanwhile, the external soundscapes that are dominated by industrial noise and vehicles during the day are replaced by noticeable quietness. However, when turning from the main roads into side streets and back alleys this silence then gives way to muted drums and bass sounds, which echo from the Hidden Agenda or nearby band rehearsal rooms and are reflected by the walls of factory buildings.

As its ambiguous and sardonic name already suggests, the Hidden Agenda operates both openly, by professionally promoting their events to the broader public, and secretly, by rendering the place invisible from the outside. In fact, no banner or street sign indicates the right direction or address of the current location. At night, people enter the venue by slipping through a small gate in the middle of a closed garage door and by taking one of the spacious industrial elevators to the second floor. Only the music itself transcends the physical boundaries of its otherwise invisible space and lends it a temporary visibility/audibility to the surrounding neighbourhood.[8]

However, the significance of the Hidden Agenda for this production lies rather in its contribution to conceptualizing the idea of the creative use of industrial buildings, which is theorized within the second pair of Lefebvre's triad, representations of space/conceived space. It mainly refers to how we make sense of and what we think about the space in which we move while simultaneously understanding this process as a mental activity that is reflected (and to a certain extent materialized) in plans, signs, models, theories, and discourses. In praxis, the formulation, interpretation, and opposition of laws and regulations play a significant role in this process and are, therefore, the main examples that are used to illustrate the duality of representations of space/conceived space hereafter.

As previously argued, the development of different urban areas in Hong Kong is primarily determined by town planning zones[9] that compartmentalize the land according to its most suitable and, therefore, strictly defined use, which can either be one- or multi-dimensional (e.g., single use or mixed use). The purpose of this ordinance is to avoid any unauthorized and sudden change in the urban landscape, such as the construction of an apartment high-rise within a country park or industrial area, where the accommodation of residents seems to be inappropriate or inexpedient. After the government has introduced its plan to rebrand the industrial part of Kwun Tong as the second Central Business District of Hong Kong—the so called "CBD[2]"—

it also adjusted the current zones by rededicating the industrial area into a business area. This adjustment entailed the legal guarantee of the conversion of industrial buildings into a commercial space, or the new construction of office towers. It also openly contested the continuation of the site-specific practices and routines of industrial workers and artists.

Therefore, the recent development needs to be understood in relation to a more fundamental change in conceiving the expediency of space in industrial areas starting from the introduction of government measures to revitalize industrial buildings in Hong Kong. Prior to this, the private and non-industrial use of former flatted factories by artists was a good opportunity for owners to generate some money from otherwise unprofitable objects. Given that most of these spaces remained private studios and were, therefore, literally kept out of sight from public discourse, an alternative use of the industrial buildings for creative work was generally tolerated, although this kind of use seemed to be in conflict with existing legislation and must have been noticed by authorities as early as 2001.[10] However, as industrial areas in Hong Kong are not conceived as merely redundant and visually unpleasant anymore,[11] a master plan such as the "CBD[2]" has been turned into an exemplary *representation* of urban redevelopment that simultaneously excludes other conceptions and spatial practices, regardless of whether its nature is industrial or creative.

Eventually, at the end of 2012, the authorities aligned the physical *representations* with their own *conceptions* and changed road signs in the district accordingly from "industrial area" to "business area," which appeared not only diametrical to the visual dominance of the rather gritty industrial buildings but also to how most of the actual users thought about and imagined their space. Unsurprisingly, local artists strongly resisted this imposed change in representation. They started a counter-campaign based on guerrilla tactics—such as graffiti and stickers—to articulate their own conception of space, proclaiming the establishment of the "Kwun Tong Art Area" instead.

At the same time, the Development Bureau of Hong Kong has been attempting to engage with local users by opening an office in Kwun Tong. The Energizing Kowloon East Office (EKEO) aims to inform the residents about the development of the project and to consult with them on proposed master plans. However, the Hidden Agenda and other creative spaces in this area had reservations about cooperation offers. Instead, the encounter even culminated in the mocking of the institution itself. When the EKEO asked locals to submit photos and artifacts to preserve and share the memories of this soon-to-be-developed neighborhood, the tenants of creative spaces responded with an Internet and leaflet campaign named Exterminating Kowloon East, using the exact same design and font of the EKEO logo. They then called for other artists and activists to submit their own photos to the original campaign only after adding a red stamp that said "Dead due to energization." Although these divergent views of conceived space eventually led to the emergence of political activism, which is undertaken by groups such as the Factory Artists Concern Group and R.I.P., a common space such as the Hidden Agenda offers a symbolic site for the contestation and negotiation of these dominant concepts.

However, it is also necessary to recognize that a space such as the Hidden Agenda has been produced by the rigid town planning zones as much as they have negated it. For instance, the strict division of residential and industrial areas renders common problems otherwise associated with nightlife venues, such as noise pollution or execution of closing hours, negligible.[12] Simultaneously, the lack of traffic and street life naturally leads to the allocation of fewer resources for monitoring that area, which is reflected in a less apparent presence of law enforcement. Nevertheless, other grounds still remain upon which the existence of the Hidden Agenda is constantly challenged, such as fire safety regulations, liquor licensing, hygiene, and public entertainment. Contradictorily, the execution of these ordinances also recognizes what the Hidden Agenda may not attempt to be: a regular live music venue in Hong Kong.

While the previous section, by offering selective descriptions, intended to demonstrate the dialectic relationship between spatial practices and representations of space (or perceived and conceived space), the third part of Lefebvre's conceptual triad goes beyond these understandings. The duality representational space/lived space is where both the perceptible and conceivable aspects of space extend to a deeper meaning that might not be adequately expressed other than symbolically—and therefore most often, though not solely, creatively or artistically (Carp, 2008: 135). Even for Lefebvre this part of the triad is "highly complex and quite peculiar, because 'culture' intervenes here" (1991 [1974]: 40).

On stickers issued for its fourth anniversary in 2013, the Hidden Agenda defined itself as a "space for live [music]" in a "Kwun Tong art district," hinting at its symbolic role for a larger community that might not necessarily address and be of any actual concern to a broader public. However, by filling the void of a mid-sized live venue for local and foreign music acts, it has not only received international media exposure,[13] but also has been given a certain extent of recognition from foreign institutions.[14] When the Hidden Agenda needed to undertake its forced relocation at the end of 2011, it decided to film the entire process and the ongoing negotiations with different authorities. The resulting documentary, *Hidden Agenda The Movie,* has since been screened on numerous occasions in Hong Kong and other cities in Southeast Asia, building up a transnational discourse that also became part of Hong Kong's official contribution to the 2012 Venice Architecture Biennale.[15] In this regard, the Hidden Agenda draws upon its quality of being what Eric Ma called (2002a) a "translocal space" that is inspired by practices, plans, and symbols from abroad (in this case, other famous music venues) and reproduces them by using the resources that are locally available.[16] This makes the space not only universally recognizable (even without knowing about its socio-spatial context), but also helps to facilitate an articulation with translocal spaces in similar situations.

However, the ongoing redevelopment plans for Kwun Tong under the broader vision that "Kowloon East should become another premier CBD of Hong Kong to support our economic growth and strengthen our global competitiveness" (EKEO, 2013a) reflect the official definition of a social and mental privilege that Lung Yingtai, in 2004, famously called "Central District Values" (Chu, 2011: 48). These values not only permeate the logic of the city's urban planning, but also create their own *representational spaces* (with a modern skyline as its materialized symbols) that manifest themselves in the *lived* experience of the users as the one "true space." The sustainability of autonomous arts clusters such as Fo Tan and Kwun Tong will therefore primarily depend on the recognition of a multiple "truth of space."

Part III: How to Lose Space

In the following part, the given context and socio-spatial analysis will be translated into a discussion of sustainability by juxtaposing the understanding of creative space in Kwun Tong with Lily Kong's study on art studios in Fo Tan (2012). However, whereas her research is conducted under a framework of cluster theory and based on an ethnographic methodology, this chapter used the smaller entity of *space* as its object for inquiry. Therefore, instead of looking at the overall sustainability of a cluster, the suggested model is adopted to evaluate a single creative space, the Hidden Agenda, from which—in its dialectic relationship with the industrial area at large and the numerous band rehearsal rooms in particular—larger implications for the creative cluster in Kwun Tong are derived.

According to Kong, the overall sustainability of a cluster is constituted by three sub-categories: economic, social, and cultural sustainability. However, their characteristics and weight might differ drastically from one place to another, which also renders Lefebvre's previously discussed conceptual triad a useful (additional) tool for analysis.

The cultural sustainability of Fo Tan is mainly tied to its industrial environment, which is beneficial for producing any kind of visual art. In fact, appropriate physical space is "a very fundamental condition for sustaining certain types of artistic work" (Kong, 2012: 186). The high ceilings, concrete floors, and spacious units that can hardly be found anywhere else than in those flatted factories are, therefore, parts of the most important arguments for painters, artisans, or sculptors to buy or rent their studios in this area. Meanwhile, the proximity to existing factories in Fo Tan offers not only convenient access to both materials and specialized craftsmanship (e.g., woodwork), but also a certain industrial "grittiness" that is often perceived as authentic and inspirational (186-7).

A creative space such as the Hidden Agenda certainly draws upon similar resources in Kwun Tong. However, although live music venues as well as rehearsal rooms can be found in other parts of Hong Kong, and their feasibility is not only bound to the specific physical characteristics of industrial buildings, their cultural sustainability must be somewhat derived from the peculiar spatial practices that are facilitated by existing zoning regulations.

While the production of visual art is of a rather introverted nature and does not affect its immediate neighbors, the Hidden Agenda and spaces nearby very much depend on the indifference and tolerance of other tenants. However, considering that none of them are (legal) residents, noise levels are presently less problematic. In addition, streets and premises remain monitored loosely within the industrial area, which also adds to the production of a subcultural imagination.

As regards social sustainability, most of the artists in Fo Tan agree that its proximity to the Fine Arts Department of the Chinese University of Hong Kong (CUHK) in Sha Tin has played a significant role for the area's social dynamics. While its influence might not be as dominant as it was previously,[17] the geographical location encourages ongoing ties between the institution and the local arts community. At the same time, the cluster also fosters interaction between different users (both industrial and creative) and a wider public, especially through local galleries and the annual Fotanian festival. However, as most of the artists prefer to work individually in their own studios, some people also seem to miss a permanent and user-friendly *common space*, such as a café or canteen that could increase the connectedness within the community (Kong, 2012: 187-91). Nevertheless, by referring to Benedict Anderson's concept of an "imagined community" (1991), Kong suggests that the mere knowledge of having like-minded people around remains one of the most important factors for motivation and inspiration of artists in Fo Tan (2012: 190).

In contrast to the individualistic art studios, the rehearsal spaces in Kwun Tong are usually collectively organized. Several bands often rent and share the same unit, which also increases the creative productivity by allowing spontaneous improvisation and collaboration between different people and across different music genres. Although this happens behind closed doors, the Hidden Agenda then offers a *common space* where they can perform or test their artistic work in front of an audience and simultaneously engage with the music of others. A similar space does appear to be partially missed in Fo Tan. However, the artists there have already established a valuable infrastructure such as the Fotanian festival, which has turned the arts cluster into a widely recognized brand. In fact, it can be argued that the Hidden Agenda serves a similar function for Kwun Tong by connecting it with a broader public, especially by inviting internationally respected bands and seeking cooperation with reputable organizations such as consulates or cultural institutes.

Other than Fo Tan, Kwun Tong is not located near any tertiary institutions, which also means that the social production of its arts cluster needs to be derived from other geographical aspects. As mentioned in the first part of this chapter, Kwun Tong is not only one of the most densely populated districts of Hong Kong, but also an area with one of the highest ratios of public housing. Nevertheless, according to the ADC survey (2010), it accommodates the largest share of Hong Kong's creative spaces in industrial buildings.

Although the survey findings do not offer a district-specific breakdown of income and educational background, the high proportion of bachelor's and postgraduate degrees among visual artists in Fo Tan certainly implies a clustering of people with relatively high social, economic, and cultural capital. Therefore, the site-specific emergence of a music cluster in Kwun Tong, in which the costs for rent are often shared by several people and credibility is not significantly linked to academic qualification, should likewise be seen as a result of the socio-economic context of the nearby residential areas and their influence on distinct spatial practices that have apparently identified music as a preferred art form.

The third and final aspect of Kong's analysis (2012) is concerned with the economic sustainability and mainly refers to how cultural clusters respond to a potential commercial development. In the case of Fo Tan, although it fulfills the most common prerequisites for commercialization, the actual effects on the popularity and affordability of spaces have remained rather low, even after the years following the revitalization measures for industrial buildings. This could primarily be attributed to its rather remote location in the New Territories, but also, again, to the strict zoning laws.

The classic gentrification process—which is commonly understood as an ongoing valorization of formerly redundant areas by artists, which eventually leads to their own displacement—appears not to have been a significant concern for local artists in Hong Kong who own or rent spaces in the industrial areas, especially compared to often cited examples from Europe (e.g., Prenzlauer Berg, Berlin) and the United States (e.g., Brooklyn, New York). While there are definitely some regions in the city that show a common gentrification process, they are located in the neighborhoods that surround the CBD, like

Sheung Wan. The industrial areas in Hong Kong, instead, have been restrained from creating their own inherent valorization cycles that otherwise could have been stimulated by their potential for cultural and creative appropriation of space, as suggested by popular creative city discourses (Landry, 2000). In the context of Hong Kong, such a process is rather imposed externally by urban redevelopment plans that, while following the logic of one "true space," shy away from questions of preservation and cultural development.

Likewise, the relative proximity of Kwun Tong to the current CBD on Hong Kong Island and large-scale real estate projects, such as the redevelopment of the former airport, accelerated the valorization of industrial premises and will eventually threaten the economic viability of the creative (and industrial) spaces, including the Hidden Agenda. However, even if the question of funding and proper legalization of the venue could be solved, once the development is underway, the increasing commercial and residential gentrification of the surrounding neighborhoods would simultaneously affect the space's current cultural (noise tolerance, enforcement of laws) and social (common space, social strata) sustainability, and a forced closure or displacement could, therefore, be inevitable.[18]

Conclusion

This chapter used the imminent redevelopment of the industrial area in Kwun Tong to trace the history, conditions, and particularities for the emergence of one of Hong Kong's largest autonomous arts clusters, in which the Hidden Agenda has been operating as an illegal music venue for more than four years. By adopting Lefebvre's conceptual triad of space, a detailed description of divergent practices, representations, and symbolisms have underlined the existing discrepancies between local users and urban planners. Kong's model of cultural, social, and economic sustainability was then used to investigate sustaining characteristics of Kwun Tong in contrast to the established cluster in Fo Tan and to argue that the wider socio-economic conditions in the surrounding neighborhoods significantly informed the emergence of an inherently *un*-elitist arts cluster.

Concurrently, the juxtaposition of Fo Tan and Kwun Tong has shown that spatial practices and representations can even vary across similar arts clusters within the same city. However, it can also be concluded that their overall sustainability depends equally on three major factors: genesis, geography, and governance (Kong, 2012: 193). Therefore, drastic changes in any of these three domains will inevitably have substantial consequences for the cluster, positive or negative.

Although the government has recently acknowledged the necessity for creative space and defined the conversion of industrial buildings as one of its policy objectives, the spatial emphasis in the analysis of the Hidden Agenda revealed an inherent lack of cultural considerations in urban planning strategies. Over the last few years, organically evolved arts clusters such as Fo Tan showed ambitions to justify their appropriation of space by building not only a platform for local artists, but also a brand that was able to articulate itself with Hong Kong's ascending gallery and auction market. Nevertheless, its (currently unthreatened) existence is rather sustained by geographical reasons than cultural ones.

However, the continuous exclusion of culture from the category of permitted activities in industrial buildings has also protected industrial areas from uncontrolled gentrification processes that would have significantly affected the affordability of spaces for artists and factory owners within a short period of time. Simultaneously, the uncertainty of the future of clusters has also generated the articulation of shared concerns across different arts sectors and facilitated the emergence of representative groups that contest dominant representations by formulating their own conceptions of space. More importantly, the Hidden Agenda has become both a symbolic site to challenge imposed redevelopment plans and a case study for survival tactics.

Therefore, its future will not only be an indicator for the sustainability of the "Kwun Tong Art Area," but also set precedence for the current limits of cultural policy within urban development projects in Hong Kong.

Notes

1. For example, Police Married Quarters (PMQ), Central Police Station (CPS), Jockey Club Creative Arts Centre (JCCAC), Comix Home Base in Mallory/Burrows Street, Cattle Depot Artist Village, Hong Kong Visual Arts Centre (VA!), Oil Street Arts Space (Oi!).
2. In this article the term "creative space" is used consistently to give a rather neutral reference to all spaces where any kind of cultural products or art forms are created. In other publications, such spaces might also be described as "alternative space," "art space," "cultural space," or "underground space."
3. In 2012, Art Basel acquired the majority of ART HK, adding an Asian edition to their existing fairs in Miami Beach and Basel.
4. "The perceived-conceived-lived triad…loses all its force if it is treated as an abstract 'model.' If it cannot grasp the concrete (as distinct from the 'immediate'), then its import is severely limited, amounting to no more than that of one ideological mediation among others."
5. For a profound analysis in the sense of Carp, many other groups and sub-groups would need to be included separately, especially industrial workers, neighbors, architects, designers, planners, lawyers, engineers, environmentalists, politicians, and so on. Given the limitation of space and resources, this chapter can only offer a small sample.
6. Given the extensive development of public transport over recent decades, the significance of the nearby ferry pier has been diminishing, but it currently continues to offer a regular schedule between Kwun Tong and Hong Kong Island.
7. Current address: 2A, Wing Fu Industrial Building, 17-15 Tai Yip Street; address in 2010/2011: 6/F, Ko Leung Industrial Building, 25 Tai Yip Street; address in 2009: 1A, Choy Lee Industry Building, 46 Chun Yip Street.

Politics and Aesthetics of Creativity

8. In his ethnographic study of a band rehearsal room in Hong Kong, Eric Ma suggested that such "underground spaces" are marked by the duality of invisibility and visibility (2002a: 138), adding a spatial dimension to Hebdige's argument that "subculture forms up in the space between surveillance and the evasion of surveillance" and that "it translates the fact of being under scrutiny into the pleasure of being watched" (1988: 35). Hebdige, however, does not particularly relate his research to a physical space (1988), and Ma, by referring to a non-public venue, sees in this duality a sort of barrier between insiders and outsiders 2002a); this chapter argues that the Hidden Agenda, instead of being a mere product of subcultural politics and emotional energies (Ma, 2002b; Collins, 1990), must be equally understood as a social construct that is constantly producing and is produced by spatial practices and representations of spaces in relation to Kwun Tong's industrial characteristics.
9. See Cap. 131 Town Planning Ordinance of Hong Kong SAR.
10. In 2001 several visual artists in Fo Tan organized their first open-studio event called Fotanian, which has since been held annually. For the rest of the year, the studios are closed to the public.
11. The EKEO (2013b) defined as one of its challenges that "the area is not particularly pleasing visually in terms of greening, open space and pedestrian environment."
12. Although certain industrial buildings are also used as living spaces, their residents usually do not pose any threat to rehearsal rooms or the Hidden Agenda, as any complaints would also reveal their own illegality.
13. Besides regional magazines, newspapers, and websites, the Hidden Agenda has also been featured by CNN Travel, China Daily, and the latest Lonely Planet Guide to Hong Kong, which described it as "the city's most visible clandestine live music venue."
14. In 2012, one concert of the annual French May festival in Hong Kong—which is presented and funded by the Consulate General of France, the Alliance Française and the Leisure and Cultural Services Department—took place in the Hidden Agenda. In December of the

same year, representatives of the Hidden Agenda were also invited to meet with Lung Yingtai, Taiwanese Minister of Culture.
15. This project entitled "Ghostwriting the Future" was officially presented by the ADC and was offered a critical reflection on the urban redevelopment of East Kowloon across various disciplines and actors.
16. Interestingly, the Hidden Agenda emphasized on its website that many foreign artists see it "as the local version of the legendary CBGB," which operated from 1973 to 2006 in Manhattan, New York.
17. During the first years, most of the artists in Fo Tan were affiliated with the Chinese University and consisted of students, graduates, and faculty staff. Later, however, more institutions started to offer their own arts programs, which diversified the composition of the cluster.
18. In the meantime, new master plans for the development of Kwun Tong consider the preservation of certain buildings in the new business area solely for creative use. However, as argued through Lefebvre's conceptual triad, a representational space such as the Hidden Agenda is simultaneously producing and is produced by its symbolic value and deeper meaning for a marginalized community. Therefore, it must be assumed that a formal recognition or even institutionalization can render the space irrelevant (Shaw, 2005: 150).

Bibliography

Arts Development Council (ADC). "Survey on the Current Status of Industrial Buildings for Arts Activities and Future Demand." December 2010. http://www.hkadc.org.hk/en/content/web.do?id=ff8081812c-2b89c5012dc5cb3d440186

Anderson, Benedict. *Imagined Communities*. New York: Verso, 1991.

Bloomberg. "Returns on Hong Kong industrial buildings 'slowing down.'" *South China Morning Post*. 17 October 2012. http://www.scmp.com/property/hong-kong-china/article/1062539/returns-hong-kong-industrial-buildings-slowing-down

Carp, Jana. "'Ground-Truthing' Representations of Social Space: Using Lefebvre's Conceptual Triad." *Journal of Planning Education and Research* 28, no. 2 (2008): 129-42.

Cartier, Carolyn. "Culture and the City: Hong Kong, 1997-2007." *The China Review* 8, no. 1 (2008): 59-83.

Chu, Stephen Yiu-Wai. "Brand Hong Kong: Asia's World City as Method?" *Visual Anthropology* 24, no. 1 (2011): 46-58.

Collins, Randall. "Stratification, Emotional Energy, and the Transient Emotions." In *Research Agendas in the Sociology of Emotions*, edited by Theodore Kemper, 41-60. New York: State University of New York Press, 1990.

Energizing Kowloon East Office (EKEO). "Vision and Mission." Homepage of the Energizing Kowloon East Office. 2013a. http://www.ekeo.gov.hk/en/vision/index.html

— "Background: Hong Kong's CBD^2." Homepage of the Energizing Kowloon East Office. 2013b. http://www.ekeo.gov.hk/en/about_ekeo/background.html

Finance Committee. "Government, Institution or Community facilities in the Kwun Tong Town Centre redevelopment—additional medical and health facilities." *Legislative Council*. February 5, 2013. http://www.legco.gov.hk/yr12-13/english/fc/pwsc/papers/p12-52e.pdf

Florida, Richard. *The Rise of the Creative Class*. New York: Basic Books, 2002.

Harvey, David. *The Condition of Postmodernity: An Enquiry into the Origins of Cultural Change*. Oxford: Blackwell, 1990.

Hebdige, Dick. *Hiding in the Light: On Images and Things*. London: Comedia, 1988.

Kong, Lily. "Improbable Art: The Creative Economy and Sustainable Cluster Development in a Hong Kong Industrial District." *Eurasian Geography and Economics* 53, no. 2 (2012): 182-96.

Ku, Agnes Shukmei. "Making Heritage in Hong Kong: A Case Study of the Central Police Station Compound." *The China Quarterly* 202 (2010): 381-99.

Landry, Charles. *The Creative City: A Toolkit for Urban Innovators*. London: Earthscan Publications, 2000.

Lefebvre, Henri. *The Production of Space*. [1974] Oxford: Blackwell, 1991.

Leung, Chun-Ying. *The 2013 Policy Address*. Hong Kong: Office of the Chief Executive. 2013. http://www.policyaddress.gov.hk/2013/eng/pdf/PA2013.pdf

Leung, Maggi W.H., and Dietrich Soyez. "Industrial Heritage: Valorising the Spatial-Temporal Dynamics of Another Hong Kong Story." *International Journal of Heritage Studies* 15, no. 1 (2009): 57-75.

Lung, Yingtai. "Cultural Policy and Civil Society: Which Possibilities does Hong Kong Have?" [In Chinese] *Journalism and Media Studies Centre, The University of Hong Kong*. 9 November 2004. http://jmsc.hku.hk/works/lung04.htm

Ma, Eric. "Translocal Spatiality." *International Journal of Cultural Studies* 5, no.2 (2002a): 131-52.

— "Emotional energy and sub-cultural politics: alternative bands in post-1997 Hong Kong." *Inter-Asia Cultural Studies* 3, no. 2 (2002b): 187-200.

Shaw, Kate. "The Place of Alternative Culture and the Politics of its Protection in Berlin, Amsterdam and Melbourne." *Planning Theory & Practice* 6, no. 2 (2005): 149–69.

South China Morning Post (SCMP). "Wong Chuk Hang's low rents attract galleries." *48 hrs Magazine.* 18 April 2013. http://www.scmp.com/magazines/48hrs/article/1213158/wong-chuk-hangs-low-rents-attract-galleries

Young, Alwyn. "A Tale of Two Cities: Factor Accumulation and Technical Change in Hong Kong and Singapore." In *NBER Macroeconomics Annual*, Vol. 7, edited by Olivier Jean Blanchard and Stanley Fischer, 13-64. Cambridge, MA: MIT Press, 1992

CHAPTER EIGHT:

The Tourist Gaze, the Invisible Cities: Cultural Heritage and Tourism in Taiwan
Yi-Chieh Lin

Introduction:
The Thousand-year-old Archaeological Site on Campus

In October 2013, an archaeological team excavated cultural relics dating back 1,500 years at the parking lot scheduled to become the new student cafeteria space in front of a public university. As soon as I heard the news, I hastened to visit the archaeological site, where the lead archaeologist, Whei-Lee Chu, who is based in the National Museum of Natural Science, was very excited to show me her findings: a beautiful jar, twenty-five centimeters wide, from the prehistoric culture Fan-zhi Yuan. This pot showed that during the prehistoric period in Taichung, Taiwan, community members shared food together. If the prehistoric people only shared food within the family unit, the pottery would be of a smaller size.

In fact, Dr. Chu told me that around the area in Taichung City (NCHU), she has discovered at least three different prehistoric cultures, dating as far back as four thousand years. The oldest culture, Niu-ma-tou, is famous for the line-patterned red, brown, and black pottery. The stone tools indicate that the prehistoric people were effective at farming, fishing, and hunting, and that they are descendants of the earliest Austronesian culture.

Figure 1: Cultural relics from the archaeological site in NCHU.

In 1975, the linguistic anthropologists Richard Shulter, Jr., and Jeffrey Marck found that the proto community for Austronesian culture, the Kadia, which were the Dai minority in southern China and the Indo-China peninsula, dated back ten thousand years, which is supported by the study of pottery of lined patterns. Approximately nine thousand years ago, certain Austronesian peoples migrated to Taiwan. The archaeologist Peter Bellwood also contended that Austronesian people are the offspring of the indigenous people who lived on the Indo-China peninsula and the present-day Yunnan and Gui-zhou provinces of China (1995).

The archaeologist Kwang-Chih Chang contended that the ancestors of the Austronesian people migrated to Taiwan by foot before the last ice age twelve thousand years ago (2010). After this ice age, the Taiwan Strait emerged and led to the evolution of Austronesian people who stayed in Taiwan. The Austronesian people in Taiwan became skilled sailors afterwards, and they sailed out to Madagascar to the west, Easter Island to the east, and became the Maori in New Zealand to the south.

The history of Austronesian migration has inspired the works of contemporary writers in Taiwan, such as Wu Ming-Yi with his novel *The Man with the Compound Eyes*. Translated into English and published by Random House, this novel is a mix of the ethnography and fiction describing the fateful encounters of different ethnic minorities of Amis and Bunun with German and Scandinavian people on the fictional island Wayuwayu in the Pacific Ocean. The author identified early anthropological works as his inspirations. For instance, in Austronesian culture, directions are referred to as "facing the sea" and "facing inland," which indicates that their earliest residential area is not an island, but rather a peninsula or continent. Therefore, Wu used these directions as common phrases used by people on Wayuwayu. The novel inspired an American with Taiwan aboriginal heritage, Tony Coolidge. His biological father, a Vietnam War veteran, abandoned him and his mother, an Atayan aborigine, after he was born. His mother remarried, another American soldier, and the family lived in different military bases around the world. Tony Coolidge obtained a degree in Advertising from the University of Texas-Austin, and never learned about his indigenous heritage until he went back to Taiwan to visit family members after his mother passed away. He wrote about his journey in the book *Voices in the Clouds*.[1] This received the attention of Aaron Hosé, a documentary producer who later helped Coolidge to finish an award-winning documentary of the same title.

In the movie, Coolidge described how he never knew about his Atayal heritage because "they have a whole history of persecution, assimilation and discrimination that has made it difficult for them to express their culture, and I was pretty shocked." Coolidge recorded the life of old women of one hundred years who wear facial tattoos, which is a tradition of aboriginal culture that is disappearing. Coolidge founded the Atayal Foundation to preserve the culture of the Atayal people, and in 2013 he founded the Austronesian Cultural and Economic Cooperation Association to promote awareness of Austronesian culture and economic cooperation between New Zealand and Taiwan.

The stories of Wu Ming-Yi and Tony Coolidge have shown that anthropology can inspire a new generation of practitioners in the Taiwanese cultural and creative industries. These practitioners use their creativity to translate scholarly works, such as ethnographies, into attractive cultural products such as films and novels to attract public attention. The revenue from the cultural products support community preservation and development of aboriginal culture, creating a sustainable cycle.

In the documentary, Coolidge spoke of his reasons for wanting to preserve the Atayal culture: "I have a little son, and I am determined that he has the opportunity to discover his roots, not from books, but from people." These words point to a common malpractice in the past endeavors of cultural preservation in Taiwan: many organizations and persons thought that once the culture is recorded as written texts or visual materials such as film, the culture will survive and thrive on its own. This is a serious misunderstanding of anthropology—anthropology aims to aid communities in preserving and sustaining lifeways and traditions based on their own decisions. Perhaps because of its colonial background, anthropology as a discipline is marginalized in Taiwan, but it has strived to become a field that nurtures a new generation of scholars that recognize our own heritage of aboriginals. In the words of Tony Coolidge, "Everybody has a little connection to their indigenous roots."

Figure 2: The stone foundation from the ruins of an original house.

The majority of the so-called "Han Chinese" in official records migrated to Taiwan during the Qing Dynasty from the coastal provinces. At the time, women were prohibited from migrating from mainland China. Hence, thousands of men who migrated to Taiwan received help from Austronesian descendants in terms of farming techniques and food. The Chinese men also married aboriginal women, sometimes through kidnapping, and settled down in Taiwan. Thus, in the local dialect of Fujianese, "wife" is said as "kanchiu," which literally means "take the hands" (of the bride and run away before the father gets you!).

Preservation Movement for Architectural Ruins (Tu-Jiao-Chuo)

Toward the end of the excavation of the archaeological site in late November, Dr. Shui-Lee Chu discovered the ruins of a house built on the thousand-year-old cultural layer of Fan-zhi-Yuan. The structure itself dated back to at least the Qing Dynasty, which is more than a century of history. The ruins continue to maintain a structure of the house's foundation, which used round stones collected from the banks of the Green River near the campus. The building technique is a common Qing way of building Tu-Jiao-Chuo, which refers to the houses that use local yellow clays with straw to create inexpensive bricks for farmers. Moreover, at the corners of these remaining foundations, archaeologists discovered bowls used by people in the Qing Dynasty, and traced their origins to certain stoves in mainland China.

Local community members also supplied additional historical data with clear evidence indicating that the Babuza, an indigenous tribe, were once very active residents at this location. During the Qing Dynasty more immigrants from China had occupied the area, also known as Dingqiaozhai by the locals. In 1885, the Qing Monarchy even planned to assign Dingqiaozhai as the official city for the provincial government in Taiwan, and started building castle walls in this area. A change of administration led to the decline of the plan, whereas Taipei emerged as an important city with the construction of railways. The official capital of Taiwan Province moved to Taipei from Dingqiaozhai in 1894, the year before Taiwan was ceded to Japan due to the 1895 Sino-Japanese War.

Dingqiaozhai was the heart of Taiwan during the Qing Dynasty. The area is famous for the rich soil, producing the highest crop yields in the present-day Taichung City. It also fostered the emergence of the famous Lin family in Wufeng and the resistance against Qing Dynasty by Shuanwen Lin. Furthermore, the area ranging from Dingqiaozhai to Wufeng was formerly occupied by the Seediq tribe, according to oral history.[2] The bloody resistance of Seediq people against Japanese colonization in Wushe was the theme of the film *Seediq Bale* (2011), which was locally translated as *Warriors of Rainbow*. This is one of the first Taiwanese movies that included aboriginals, who were not professional actors, playing major roles, speaking and singing songs in their own languages. The movie has contributed to new cultural tourism, encouraging visits to Seediq villages, now based in the Hsinyi District of Nantou County, near the Sun Moon Lake. The Hui-Sun Forest near the Sun Moon Lake used to be a Seediq forest until the Japanese occupation, during which time it was assigned to a public university as an "experimental forest" for use as a coffee plantation and other research purposes.

Figure 3: An old Seediq woman, nicknamed Chang mama, who starred in *Seediq Bale*, telling life stories and showing costumes to a group of college students from Taipei.

The Seediq tribe people also suspect that one of the cultural layers found in urban Tiachung involves the migration history of cultural heritage of the Austronesia culture. This view will require additional research on material culture based on the oral testimonies of the seniors in the Seediq tribe, and the cultural relics found in the excavations in NCHU and even in central Taiwan.

When Japanese colonizers came to Taiwan in 1896, they also invited a Scottish engineer, William Burton, to design a sewage system for the island's cities. Around the ruins of Tu-Jiao-Chuo in Taichung city, a hydraulic engineering scholar, Dr. Shu-Chin Chen, identified the sewage system, built as a heritage from the Japanese period. Dr. Chen noted that plastic tubes were installed in the modern period in the latter half of the twentieth century. Numerous artifacts discovered here include glass tubes for laboratory experiments during the Japanese Period, as well as teacups used by Japanese professors nearly one hundred years ago. During the Japanese colonial period, public universities were installed as research bases for Japanese imperialists to develop their colonial economy. In the school archives one can read the digital and actual theses written by the Japanese on the development plans for tea and sugar cane plantations, as well as rice, banana, and forestry projects to produce exports and compete with other major imperial countries at the time.

Figure 4: Broken pieces of concrete tubes of the former sewage system from the Japanese Period

In central Taiwan indigenous people continued to receive minimal support in terms of community building or cultural preservation, even though the Seediq people possess a rich culture that survived after the Japanese occupation. The Seediq people are very warm and generous—I even met a Japanese professor who became "adopted" by the tribe people by way of receiving a formal Seediq name. The younger generation of Seediq children have been working hard to learn their own language, songs, and dances, but additional research is required in terms of preservation of the traditional lifeways: food tradition, witch doctors, preservation of the original residences, and histories of migration routes.

That said, I submitted a formal application to register the ruins on the public university campus as a formal cultural heritage at the Taichung Municipal Government. The city government officials invited top professional archaeologists to investigate the archaeological ruins.

Even the scholars in the areas of Creative and Cultural Industries, Architecture, and Tourism in nearby universities have joined the preservation movements. Professor Eaton Kuo in the Creative and Cultural Industries Department of National Taichung University of Education has applied for funding from a Taiwanese non-governmental organization (NGO) to build an "outdoor museum" in Taichung City. His idea is that the archaeological ruins can be preserved as one element of the outdoor museum to be incorporated into the spatial design for a new student café. This new cafeteria can be a multifunctional space serving organic food supplied by local farmers. The café can also serve as a kitchen for students to exchange new ideas and innovate new, healthy recipes in reaction to food safety problems. During off-peak hours, the cafeteria can become a space for students to brainstorm ideas of social design and innovations. Lastly, parts of the space can be used for galleries to exhibit cultural relics and local natural history, such as entomology, water preservation, forestry, and diet. Students, including college students and aboriginal teenagers, can be recruited as guides for the gallery exhibitions, which, in Canada, has proven to be an effective way to stimulate student interest and to increase the pride and understanding of one's own culture in museums (Rowan, 2012).

Food as Cultural Heritage and Weapons of the Poor
The slow food movement vis-à-vis memories of *terroir*

> "To change the world, you have to change the menu first." *Gigi Padovani, a slow food movement participant*

Food is nutrition for human bodies; food also embodies social identities. Food is a lens by which to examine social relations, social changes, family and kinship, class and consumption, gender ideologies, cultural symbolisms, risks, and disasters brought by globalization in the case of many food safety issues (Watson and Caldwell, 2004). Food is also seen as a new type of cultural heritage, as recently recognized by UNESCO. According to the anthropologist Sidney Cheung, food heritage should be understood by three factors: its production system, its distribution channels, and family and traditional recipes (2013). The anthropologist Joyce Yeh has contended that food can also be used as weapons by aboriginal tribes in eastern Taiwan to safeguard the traditional knowledge of aboriginals in aboriginal restaurants (2009). The anthropologist Hung-Yu Ru has conducted research to preserve local food knowledge in northern Taiwan. Since 2014, Seediq tribe has initiated a new project to preserve the local food culture of the aboriginals in central Taiwan. To preserve food heritage also means to preserve agricultural heritage, and defend biodiversity and cultural diversity, in the spirit of the slow food movement.

In the following case study of the restoration of cultural space in downtown Taichung, we can see that *creative design* can unite the pleasure of food with responsibility, sustainability and harmony with nature, history, and sensations of visual experiences in the spatial attention to aesthetical details. These factors have led to a tourist boom and promoted a new model of cultural economy based on local knowledge.

This story began with the two owners of the Dawncake Company, who are industrial designers-turned-pastry shop owners. They renovated an eye clinic and turned it into a pastry and ice cream shop. The eye clinic housed a hundred beds during its heyday in the 1930s and early 1940s, and was one of the largest hospitals in Taiwan, located in the intersection of the Zhongshan Road and Liuchuan East Road, next to the Zhongshan Bridge across the Green River, built in 1908.

The eye clinic belonged to Miyahara Takaguma, an optometrist trained in Berlin University and Vienna University, and who received a PhD in Medical Sciences from Tokyo University. Dr. Miyahara was also president of a high school and a right-wing councilor in Taichung. Because of conflicts of interest among the right-wing advocates in Taiwan, Dr. Miyahara allied himself with the local Taiwanese and became an active advocate for the rights of the local Taiwanese population. He founded the East Asia Co-Prosperity Council, where Taiwanese people enjoy the same rights of speech as the Japanese. His sympathy towards the locals led to physical attacks of Dr. Miyahara by Japanese police, and he was deported back to Japan after the Second World War (Wang, 2013).

Figure 5: Miyahara Clinic, after renovation

The clinic was turned into a public hospital, and in 1958 was sold to a merchant, Mr. Chang, in exchange for capital to build a new public hospital elsewhere. However, Mr. Chang had failed to dismiss illegal occupants of the buildings, who did not leave until the 1999 earthquake. In addition, in 2008 a typhoon devastated the main hall and second floor of the building. Mr. Chang sold the property to the Dawncake Group. The executives, Shu-Feng Lai and her husband, have backgrounds in industrial design, and began running a pastry business based in Taichung. Archi Su, one of the three architects commissioned to renovate the Miyahara Clinic, grew up in the area around Chien-Kuo Market. In a magazine interview, Su expressed nostalgia for downtown Taichung. According to the Code of "Rewards for the Renewal Volume of Urban Regeneration," the renovated building could be as tall as thirteen stories. However, the architects recognized the fact that only the ground floor would interest visitors, and the renovated building thus maintained a three-story structure. The executives of the Dawncake Group presented only one goal for the architects: "This building should help to revitalize the downtown area." The city government provided assistance for renovation and construction licenses. The wooden tiles of the second floor were reused to become display stacks and decorations.

Figure 6: The interior design of Miyahara Eye Clinic, which uses traditional Chinese medicine shelves and eye tables as elements to create a nostalgic atmosphere in the shop.

Politics and Aesthetics of Creativity

Figure 7: The interior design of Miyahara Clinic Pastry Shop, which includes elements in the Art Deco style that was popular during the Japanese Occupation.

To fully utilize the theory of "narrative design," the interior design of the Miyahara Clinic "pastry shop" also includes decorations of shelves that look like traditional Chinese herbs stores, as well as Art Deco glass ceilings and marble floors. In the shops, one can also obtain free postcards printed with a local poet's appreciation of agriculture, as well as photographs of the pineapple farmers whose produce is used in the pastries. The vision table is also a design that evokes a sense of history and social concern for food safety and preserving local agriculture and culture.

Ms. Yu, an informant who has worked for the Dawncake Shop, said that she was determined to join the business after college graduation because she likes the atmosphere of the shop. In the Dawncake Group, every staff member is able to share their ideas about product design or store design, regardless of their academic background.

Another historical building, the No. Four Credit Union Building, was renovated by the same corporation and opened in August 2013. It also quickly became a popular spot; people wait for twenty minutes to get ice cream, drinks, or other food, taking photographs as they wait. It is located across from the Downtown Revitalization Project Office.

Figure 8: Taichung No. Four Credit Union Building, after renovation.

An approximately thirty-minute walk across Taichung Park is the Miyahara Takaguma, which was used as the city mayor's house for decades and is now an open public space for exhibitions, and restaurant space run by the city government. However, the food menu lacks creativity and the exhibition space is often closed. This place's story should be contextualized in the macro-narrative design of urban history. Next to the Miyahara residence is the first radio station and stadium. These places would attract more visitors if they were more pedestrian-friendly, with improved signs and longer opening hours, possibly also adding evocative illustrations of the history, and even a place for children to play.

Apart from pastry shops, Taichung is also famous for supporting independent, fair-trade coffee shops, which has fostered the growth of third-space and civic society. Certain coffee shops also support local NGOs by advertising performances or providing venues for local artists to showcase their works. However, the city government fined one of the coffee shops for hosting live jazz music without a license about three years ago. Since then, jazz fans lamented the demise of a place for gathering and discussion. Less bureaucracy would be another key to the success of the "Creative Taichung Project."

The Urban Fabric of Taichung as the Source of Innovation

The name Taichung consists of two characters: Tai, which refers to Taiwan, and "Chung" means "impartiality" or the concept of "neither too much, nor too little." Some believe that the nickname "City of Culture" has much to do with the active cultural movements based on the Central Bookstore, which was established in 1927 in Taichung. Led by Mr. Hsien-Tang Lin (Lîm Hiàn-Tông in Taiwanese) from Taichung, the following organizations were established in the 1920s: Taiwan Culture Association, Taiwan Local Autonomy Alliance, Taiwan People's Party, and Taiwan Farmer's Cooperative. Hsien-Tang Lin believed in a non-violent approach in the civil rights movement against the Japanese colonizers. His residence, the Lin Family Garden in Wufeng, Taiwan, was the inspiration of a CCTV drama, *Chang-hai-bai-nien*, which describes the Lin family's immigration to Taiwan and their history from 1787 to 1885.

However, as the city grew from forty thousand in 1900 to 2.69 million in 2013 (Taichung City Government Statistics),[3] certain citizens considered old houses as hindrances to development. In 2012, a group of landowners and developers entered an empty, eighty-two-year-old Chinese style three-section compound house (*san-ho-yuan* in Chinese) at two o'clock in the morning and used excavators to destroy part of the building. After the police investigation and NGO protests, the city government designated the compound house, Rui-Cheng-Tang, as a historical building and asked the developers who destroyed the building to pay for the rebuilding expenses at approximately US$30,000. This incident shows that different groups of people have different opinions about aesthetics of urban space. Different views on historical buildings have reflected diverse ideas of the local identity of the place. As Claire Parin points out, the very idea of the local identity of a place seems to be essentially subjective, intrinsically linked to whoever describes it, and unable to be dissociated from the cultural context from which it emerges (2007). Such a subjective, even sentimental position is somewhat paradoxical given the objective orientation that legitimates much contemporary discourse. Such a position recalls the ambiguity noted by Alois Riegl in 1903, at the beginning of the twentieth century, concerning the concept of "ancientness," a quality that has proved particularly effective in promoting the preservation and conservation of material elements of cultural heritage able to be identified by specialists as worthy of preservation (Parin, 2007:15).

A colleague, Professor Chang-Yian Lee, has pointed out that "to construct a new urban aesthetics is necessary to recover the 'soul of the city' and to improve the soft power of Taichung City. The stereotypes of Taichung include housing development booms, luxury houses of conspicuous consumption, low occupancy rates, and active night clubs. The images of Taichung Park and Pavilion are old aesthetic sense. The new urban aesthetics should include revitalizing cultural heritage by the partnership of public sector and private sector".[4]

The Sanmin Road, nicknamed the "Bridal Salon Street" in Taichung, is one of the shopping areas that provide an important service in the old downtown. The creative styles of bridal glamour shots have made this industry appealing to international customers from Korea, Japan, Southeast Asia, Hong Kong, and Macao. It is also the birthplace of Taiwanese style bridal rental and glamour photography. The first photography studio in Taichung was established in the early 1920s by Cao Lin. Although Lin was a street cleaner in the downtown area, he later became an apprentice to a Japanese photographer. After his mentor returned to Japan, Cao Lin assumed his studio, and his son married the granddaughter of a Member of Board of Director of Changhua Bank, the first bank run by Taiwanese citizens during the Japanese colonial period. However, the Lin Photography Studio became bankrupt. The wife of Cao Lin's son then went to Japan to learn hairstyling, clothes tailoring, fashion, and cosmetics. Upon her return, she helped her husband to open a new photography studio with Western-style bridal dress rentals, hair design, and make-up services for newlyweds. Other people soon imitated this style of operation, resulting in the widespread bridal rental businesses (for more information see Bonnie Adrian's book, *Framing the Bride*, 2003).

In the 1980s and 1990s, the clustering of bridal salon and photography businesses promoted clustering effects of consumer bases in central Taiwan and overseas. Increasing competition, environmental degeneration, and excessive commercialization drove certain businesses to the outskirts of the city, partly contributing to urban sprawl. The degeneration of downtown and failures of revitalization projects in the past decade were due to the landowners not being active participants in these projects. In order to preserve the collective memories, we suggest that local government pay more attention to the urban fabric and urban history, highlighting tangible and intangible heritage in place-making. Instead of building new department stores to house all bridal salon businesses, as certain local residents suggested, we argue that revitalizing obsolete buildings is an improved approach. This will maintain pedestrianism and introduce a "soft edge" to the street, with shops lined up with transparent façades, large windows, and numerous goods on display. Sanmin Road is an example of this approach, where shops provide reasons to slow down or even stop at the Second Public Market for traditional food. Several shops attract visitors from Hong Kong and mainland China. Bubble tea originated in Taichung, and the shop where the beverage was created is now frequented by many local residents and tourists, owing to its Chinese flower arrangements, wooden carved screens, and excellent food and drinks.

Jan Gehl defines an edge as the place where building and city meet, such as the lower floors of buildings (2010: 75). These are the frontages that are seen and experienced up close, and therefore, intensely. He said:

> The edges of a city limit the visual field and define individual space. Edges make a vital contribution to spatial experience and to the awareness of individual space as a place. Just as the walls of a home support activities and communicate a sense of well being, the city's edges offer a feeling of organization, comfort and security. We recognize the space with no edges or weak edges from many urban squares with heavily trafficked roads on all four sides. Their function is considerably more impoverished than the city space where life is directly reinforced by one or more attractive edges.

Our Dawncake informant, Ms. Yu, notes that customer behaviors are different in the various Dawncake branches (downtown, the Calligraphy

Greenway, and Taiwan Avenue). She said that customers in the downtown and Calligraphy Greenway shops tend to stay longer and will also walk to other shops nearby. Customers who visit the Taiwan Avenue branch are always hurrying, and so the shop even added a drive-through for the customers.

However, the area also faces core issues including aging population, consumers' preferences of using Internet technology to sort out bridal salon choices, and the ownership of land rights. According to one informant who grew up in downtown Taichung during the 1950s and '60s, many of her relatives and neighbors have migrated to United States or moved to the periphery of the area. The city government commissioned the National Taichung University of Education to plan an "electronics shopping district," similar to the Akihabara in Tokyo, but the customer base continues to decline. Another colleague, Professor Jui-Pi Su from Tung Hai University, who has started a new initiative of urban revitalization, commented on his project in collaboration with the city government: "It is important to motivate the residents and private sector instead of using the top-down approach." Hence, in the Downtown Revitalization initiative, the DRF Goodot Village Project, [5] the members not only conduct an investigation of local history but also mediate between land- and house-owners and tenants, applying creative methods to increase the city population in line with a limited budget. The Goodot Village Project Office is located in a beautiful old building that was once a bank office. They also found that these buildings offer texture, solid materials, and a wealth of details, such as Gothic concrete tiles in the ceiling, terrazzos, and window sills, as well as spiral staircases, which are scarce in modern architecture. The second-floor windows provide a good vantage point to observe the backstreets of downtown. Based on project investigations, Taichung's grid city plan was successful—several natural streams, where leisure activities took place, flow across the city, and shade trees provided protection for walking pedestrians during the hot summers. The designers, a group of young architects, revitalized the community area by mediating between young people in the creative industries and the local community, leading to people staying in the neighborhood. For instance, people who attend cram schools in this area would stay after class. The Goodot Village also set up exhibitions of photographs showing the history of downtown Taichung, and recorded stories of professions and their century-old history: Chinese herbs, tailors, stationery stores, pastry bakers, bookstores, noodle stands in the markets, eye clinics, and so forth. This information was published in hard copy as well as on social network sites to get comments, and also in department stores and outdoor museums (see photos). Encouraging pedestrianism is a main goal in the project, and the team also encouraged ideas to "green the old buildings" by applying horticulture to obsolete department stores. The new tenants of the old buildings used crowdfunding websites to raise capital to renovate the buildings and receive comments for their projects. One turned an attorney's office into a shared office space for creative industries, and another turned an old house into a bookstore and gallery space.

Figure 9: DRF Goodot Village Project is located on the second floor of the building.

The Tourist Gaze, the Invisible Cities

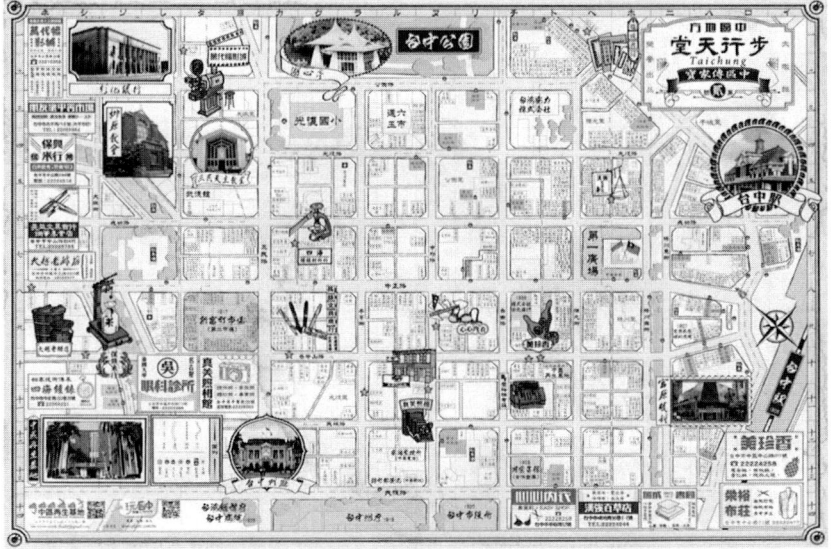

Figure 10: This is one of the publications of the Goodot Village Project, which highlights the interesting historical architecture and different old businesses that have remained in the area for decades.

Figure 11: This is a pedestrian map to illustrate young entrepreneurship in the downtown area.

Figure 12: Many beautiful historical buildings remain. This is Dr. Wu's Eye Clinic, which is still in practice.

In September and October 2013, Taichung hosted a series of Creative City Events, including an outdoor exhibition to highlight ten "creative clusters" of different characteristics, including the work of artists and designers. These districts are listed as follows:

Features	District Name
Design, bookstores	Calligraphy Greenway
Lacquerware, pastries	Feng-yuan
Arts, music	Tung-Hai University Area
Historic buildings	Shuan-Shi Water Reserve
Young entrepreneurship, cultural and creative industries, historic buildings	Downtown (Central District)
Innovative market food, shopping	Feng-Chia University Area

Horse range, bicycle lane, saxophone	Holi
Cinema, 1960s garden city layout (Guangfu New Village)	Wufeng
Food, handmade hats and crafts of Cyperus malaccensis	Dajia
Mushrooms production base, flower gardens, and coffee plantations	Shinshe

    This list shows the ambition to promote tourism by using the label "creative city" to highlight the potential of leisure businesses in these regions. However, the list is not complete. For instance, Wufeng is also home of the Taiwan Symphony Orchestra and has the potential to develop a local music industry. Most of these places remain unknown to international tourists, and they face other challenges. Many "garden houses" of the Guangfu New Village are unoccupied, making a visit to the place a boring experience without interactions with the people. Despite a visible foreign population in Taichung, including the Japanese Women's Association, an American school, and a Japanese school, openness to multiculturalism is not shown in the "creative city" outline. Compared to Tainan and Taipei, the creative spaces for artists and designers (such as galleries, studios, or coffee shops) are more scattered in different areas of Taichung, and construction of mass transportation connecting different creative districts is required. Taichung's Creative City Plan could succeed by promoting the cultural and creative businesses along with other cities in and out of Taiwan, saving the promotion budget and increasing awareness. The outdoor museum on the Calligraphy Greenway (Figure 13) is scheduled to be demolished in order to construct high-rise buildings. To promote more outdoor museums like this, certain NGOs donated millions of dollars to establish a new outdoor museum in Taichung.

Politics and Aesthetics of Creativity

Figure 13: The Creative City Outdoor Exhibition. Each house displays a map of a creative district and its featured objects.

The Tourist Gaze, the Invisible Cities

Figure 14: Another historical building recently renovated. The landlord posted an advertisement looking for tenants.

The district's main specialty businesses remain the bridal salon industries as well as pastry shops. The city government hosts competitions to rank the best bakery every year, and tourists from mainland China are likely to visit and buy pastries as souvenirs. Taking advantage of a new trend in marketing, in 2013 the city government hosted a student competition on the Internet using microfilm to tell the stories of the pastry shops to attract more attention.[6]

Taichung, a city of 2.9 million, is a mega-city with a recent amalgamation to merge the metropolitan area of industry-based economy with the surrounding county of an agriculture-based economy. Taichung partnered with telecom companies to create thousands of WiFi hotspots, fiber-based broadband, and 4G WiMax wireless signals to reach 90 per cent of the population. The result has been an economy driven by information and communications technology, allowing a network of 1,500 precision machinery makers and tens of thousands of server message block suppliers to produce a US$30 billion output. Taichung City has also made up a "Low Carbon City Project" to cut down carbon dioxide emissions. Covering an area of 2,215 square kilometers, Taichung has eighteen colleges, thirty-one libraries, forty-two bike lanes (totaling 262 kilometers) and two hundred-twenty parks. The project is modeled after the Swedish city of Malmö, aiming to transform an industrial area into an eco community: the Taichung Gateway Park City. The City also encourages vegetarian diets by hosting seminars and promoting vegetarian diets in public sectors and schools.

The Oldest Graffiti Artist in the World—The Story of Rainbow Village

Richard Florida coined the term "creative class," which loosely includes artists, cultural creatives, student professionals—designers, programmers, physicians, attorneys, accountants, professors, financial specialists (2011). Florida's economic model is driven by the forces of 3T: Technology, Talent, and Tolerance (an openness to all kinds of people, regardless of gender, race, nationality, and sexual orientation), which are key to economic vitality. Florida emphasizes that building a "people climate" is important to attract the diverse human talents who could contribute to true prosperity. To improve the "people's climate" is as simple as improving neighborhood conditions with small investments in everything from parks and bike paths to street-level culture that would improve people's everyday lives, thus improving the underlying quality of place. These physical infrastructures signal and stimulate a community that is open, energized, and diverse.

In Florida's logic of creative economy, human resources, and talents are the main driving force that could fully harness creative energies of everyone to build a society that acknowledges and nurtures the creativity of every human being. The Civic Affairs Bureau of Taichung City has yet to conduct a statistical analysis about the creative class percentage. I used the database in household registration system to find that approximately 17 percent of the registered population of Taichung possess college degrees or above. However, this figure does not account for the many workers in the city who might be registered in another city or county for various reasons (e.g., local identity, tax purposes, incentives, or a lack of motivation to record the change). However, Taichung City has outperformed Toronto and Stratford and won the title of Intelligent

Community of 2013, awarded by the Intelligent Community Forum (ICF) in New York City (Jackson, 2013).

Florida suggests that every person has creativity of different degrees that deserves to be utilized. This argument is less emphasized in his book, but here I would like to use the case of the Rainbow Village amateur artist to show that creativity often manifests in the everyday life of the most ordinary people. The Rainbow Village's name comes from graffiti paintings by a ninety-year-old retired soldier from Hong Kong, Yong-Fu Huang. The graffiti, although technically illegal, became a rallying cry for the village's inhabitants when the government made plans to tear down the village to make room for public housing. Students began petitioning on social networking sites, and the village received widespread attention from the media and politicians in 2011. The Rainbow Village is now permanently preserved, although nearby military villages have been torn down for new construction despite the fact that some of them began imitating the graffiti of Rainbow Village in attempts to preserve their spaces. Successful commodification of this new urban landscape has occurred, such as the uses of these images in department stores and on mugs for sale in bakeries.

The mediascape of the "military village" is both medium and message. In the case of the Rainbow Village, graffiti, social media, and films about the place are both media representations and "weapons of the weak" in the hands of preservationists (Scott, 1985). The art of resistance represents a local reversal of "the global hierarchy of values" (Herzfeld, 2004). This form of resistance also brings to the fore a form of social embarrassment in an effort to protect the inhabitants' social rights to the place and prevent their deterritorialization. Thus, the movement is also a form of "cultural intimacy" (Herzfeld, 1997).

Figure 15: Young street artists playing on the weekends at Rainbow Village, interacting with audience and visitors.

Graffiti artists have received increasing attention by society and have been recognized by official art circles. The National Museum of Fine Arts in Taichung has hosted exhibitions of the New Taiwan Wall Painting Group.⁷ Candy Bird is another popular graffiti artist who addresses social issues. One high school student told me that his favorite graffiti artist is Virus No. 6. In Tainan, graffiti art has also been important in urban revitalization projects by young people, resulting into the emergence of the "backstreet culture" (xianglong wenhua), which is the new buzzword in the Taiwan cultural and creative industries.

The Tourist Gaze, the Invisible Cities

Figure 16: Candy Bird graffiti, taken in Taipei by author

The *Finale*: On the "Unofficial Cultural and Creative Industry" and Urban Gentrification Process in Taiwan

In the last part of this chapter I would like to address the issue of gentrification and its negative impact on the development of the cultural and creative industries in Taiwan. Across from the National Fine Arts Museum, avant-garde galleries such as Z Space and other independent theaters and cafés have brought art into Zhongxin Farmer's Market. It is a place where avant-garde artists mingle with other tenants—mostly senior citizens, homemakers from Southeast Asian countries, and children—as a free-style community. The gallery, Z Space, is supported by a group of sponsors who pay NT$1,000 monthly for the rent. It provides a no-fee place for young artists to showcase their works and ideas. On the second floor, the designer for this place, Xiaoyu, created a narrow toilet corridor that terminates at a "toilet study room," a large creative space displaying books and artwork. In the "penthouse" of Z Space, one can see the front of the National Fine Arts Museum while sitting on the bar stools, chatting with the curator and other visitors. Most of the visitors of the avant-garde gallery are young people, including curators from Taipei and those who wish to be Taichungese. Many of the avant-garde artists I spoke to in interviews are from central Taiwan. One of the leaders is an expatriate who retired from his job in New York, who has a passion for arts and social justice.

Figure 17: A café/gallery in Zhongxin Market.

The Tourist Gaze, the Invisible Cities

Figure 18: An avant-garde piece at Z Space.

Figure 19: Looking over the National Fine Arts Museum from above Zhongxin Market.

The avant-garde artists, however, are worried about the forces of gentrification. The rents of the majority of spaces in the markets have increased since it became a famous and popular place for tourists, as well as for real estate developers. Property prices, such as for apartments, have surged more than 300 percent in ten years for new houses. Real estate developers attempted to tear down the Zhongxin Market by offering a price of NT$1 million per ping (approximately three square meters) to the landlords. The artists told the landlords to lease the space for another three years and said that "the land price will be even higher." By speaking in the logic of capitalism, these artists hoped to be able to change the values and mindsets of the people in the city with time.

The city government also hosts festivals for the immigrants from Southeast Asia, such as the Songkran Festival, a traditional Thai festival, in order to increase the understanding of different populations in the city.

Figure 20: The curator (middle) with two founding artists chatting in the penthouse of Z Space.

Figure 21: The "development" and gentrification of the area, which is turning the skyline of Taichung from green shades of trees to high-rise, kitsch, concrete buildings.

Over the past decade, Taiwan has been split by political interests as well as gentrification processes by different land development plans. The "development" and gentrification of the area has transformed the skyline of Taichung from the green shades of trees to high-rise, kitsch, concrete buildings that are symbols of neo-liberalism. The time is ripe to reconcile the spirit of the Austronesian heritage; that is, to build harmony based on the core concept of sharing, courage for exploration, and cherishing tradition. Sharing is the type of message that could be learned from the artifacts of the archaeological excavation. Sharing is key for the development of cultural and creative industries, and preserving the archaeological site will forever preserve this spirit of sharing and the wisdom from our ancestors for human survival, overcoming climate change, and the shifts brought by neo-liberalism.

Notes

1. A link to the preview of the documentary *Voices in the Cloud* on YouTube http://www.youtube.com/watch?v=atnp0HhIXlc.
2. Before the eighteenth century, the space of Taichung was inhabited by aboriginal tribes—the Pazeh, Papora, Babuza, and Seediq. The Han Chinese immigrants established forty farming villages and three market streets in the eighteenth century.
3. http://demographics.taichung.gov.tw/Demographic/Web/TCCReport01.aspx
4. http://ctee.com.tw/news/view.aspx?newsid=16990&cat=7
5. https://www.facebook.com/GoodotVillage
6. https://www.facebook.com/2013cakemovie?ref=hl
7. They also have a Facebook Page to browse their previous works at https://www.facebook.com/FWPG88

Bibliography

Adrian, Bonnie. *Framing the Bride: Globalizing Beauty and Romance in Taiwan's Bridal Industry*. Berkeley: University of California Press, 2003.

Bellwood, Peter, James J. Fox, and Canberra: ANU E Press, 1995.

Certeau, M. de. *The Practice of Everyday Life*. Translated by S. Randall. Berkeley: University of California Press, 1984.

Chang, Kwang-Chih. *Fengpitou, Tapenkeng, and the Prehistory of Taiwan*. New Haven: Yale University Press, 2010.Chen, Jian-Ni. "Interpreting the collective memories of 'bridal salon street' in Taichung in the Perspective of Cultural Production of Consumptionscape." Master's thesis, National Chung Hsing University, 2012.

Cheung, Sidney. "From Foodways to Intangible Heritage: A Case Study of Chinese Culinary Resource, Retail and Recipe in Hong Kong." *International Journal of Heritage Studies* 19, no. 4 (2013): 353-64.

Fogelson, Robert. *Downtown: Its Rise and Fall, 1880-1950*. New Haven: Yale University Press, 2001.

Gehl, Jan. *Cities for People*. Washington, DC: Island Press, 2010.Harvey, David. *Spaces of Capital: Toward a Critical Geography*. New York and London: Routledge, 2001.

Herzfeld, Michael. *The Body Impolitic: Artisans and Artifice in the Global Hierarchy of Value*. Chicago: University of Chicago Press, 2004.

— *Cultural Intimacy: Social Poetics in the Nation-State*. New York: Routledge, 1997.Jackson, Brian. "Toronto misses out on Intelligent Community Title to Taichung, Taiwan." *IT Business Canada*. 10 June 2013. http://www.itbusiness.ca/news/toronto-misses-out-on-intelligent-community-title-to-taichung-taiwan/35963

Jacobs, Jane. *The Death and Life of Great American Cities.* [Reissue] Vintage, 1992.

Kwok, Jackie. *Production of Space in East Asian Cities.* [In Chinese] Taipei: Garden City Culture Press, 2011.

Landry, Charles. *The Creative City: A Toolkit for Urban Innovators.* London: Earthscan, 2008.

Parin, Claire. "Introduction: New Ways to Read Difference." In *Cross-cultural Urban Design: Global or Local Practice,* edited by Catherin Bull, Davisi Boontharm, Claire Parin, Darko Radovic, and Guy Tapie, 15-23. London and New York: Routledge, 2007.

Redfield, Robert. *The Little Community and Peasant Society and Culture.* Chicago: The University of Chicago Press, 1960.

Robertson, Jennifer. *Native and Newcomer: Making and Remaking A Japanese City.* Berkeley: University of California Press, 1991.

Rowan, Madeline. *Indigenous Teenage Interpreters in Museums and Public Education.* Shawnigan Lake, BC: Diamond River Books, 2012.

Scott, James. *Weapons of the Weak.* New Haven: Yale University Press, 1985.

Shutler, Richard, Jr., and Jeffrey C. Marck. "On the dispersal of the Austronesian horticulturalists." *Archaeology and Physical Anthropology in Oceania* 10, no. 2 (1975): 81-113.

Wang, Pai-Ren. "Eye Clinic that Sells Ice Cream: Miyahara Takeguma and his street house boom." *Taichung: Ming Dao Arts and Literature Journal* (2013):105-8.

Watson, James, and Melissa Caldwell, eds. *The Cultural Politics of Food and Eating.* Malden, MA: Blackwell Publishing, 2004.

Yeh, Joyce. "From Food to Weapon: A Socio-Cultural Analysis of Indigenous Restaurants in Hualien." *Studies on Humanity and Ecology in Taiwan* 11, no. 1 (January 2009): 29-60.